ABOUT THE COVER

[Tropical Forest]

NEW BRITAIN, NEW GUINEA

New Britain is an island near the island country of New Guinea in the Pacific Ocean. New Britain has volcanoes, caves, rivers, and a dense tropical forest with some of the tallest tropical trees in the world. The island of New Britain is also one of the least researched and least explored places on Earth.

The island is thought to have 16,000 different species of plants. It also has wide variety of animal life including marsupials, birds of paradise, and the world's longest lizard, the Papua monitor.

One of the animals that can be found in the vast tropical forests of New Britain is the endangered long-beaked echidna. The echidna is one of the two types of mammals that lay eggs!

NATIONAL GEOGRAPHIC SCIENCE

LIFE SCIENCE

School Publishing

PROGRAM AUTHORS

Randy Bell, Ph.D.
Malcolm B. Butler, Ph.D.
Kathy Cabe Trundle, Ph.D.
Judith S. Lederman, Ph.D.
David W. Moore, Ph.D.

Program Authors

RANDY BELL, PH.D.
Associate Professor of Science Education,
University of Virginia, Charlottesville, Virginia
SCIENCE

MALCOLM B. BUTLER, PH.D.
Associate Professor of Science Education,
University of South Florida, St. Petersburg, Florida
SCIENCE

KATHY CABE TRUNDLE, PH.D.
Associate Professor of Early Childhood Science
Education, The School of Teaching and Learning,
The Ohio State University, Columbus, Ohio
SCIENCE

JUDITH SWEENEY LEDERMAN, PH.D.
Director of Teacher Education,
Associate Professor of Science Education,
Department of Mathematics and Science Education,
Illinois Institute of Technology, Chicago, Illinois
SCIENCE

DAVID W. MOORE, PH.D.
Professor of Education,
College of Teacher Education and Leadership,
Arizona State University, Tempe, Arizona
LITERACY

Program Reviewers

Amani Abuhabsah
Teacher
Dawes Elementary
Chicago, IL

Dr. Maria Aida Alanis
Science Coordinator
Austin Independent School District
Austin, TX

Jamillah Bakr
Science Mentor Teacher
Cambridge Public Schools
Cambridge, MA

Gwendolyn Battle-Lavert
Assistant Professor of Education
Indiana Wesleyan University
Marion, IN

Carmen Beadles
Retired Science Instructional Coach
Dallas Independent School District
Dallas, TX

Andrea Blake-Garrett, Ed.D.
Science Educational Consultant
Newark, NJ

Lori Bowen
Science Specialist
Fayette County Schools
Lexington, KY

Pamela Breitberg
Lead Science Teacher
Zapata Academy
Chicago, IL

Program Reviewers continued on page iv.

Acknowledgments
Grateful acknowledgment is given to the authors, artists, photographers, museums, publishers, and agents for permission to reprint copyrighted material. Every effort has been made to secure the appropriate permission. If any omissions have been made or if corrections are required, please contact the Publisher.

Illustrator Credits
All illustrations by Precision Graphics.
All maps by Mapping Specialists.

Photographic Credits
Front Cover Stephen Alvarez/National Geographic Image Collection.

Credits continue on page EM16.

Neither the Publisher nor the authors shall be liable for any damage that may be caused or sustained or result from conducting any of the activities in this publication without specifically following instructions, undertaking the activities without proper supervision, or failing to comply with the cautions contained herein.

The National Geographic Society
John M. Fahey, Jr.,
President & Chief Executive Officer

Gilbert M. Grosvenor,
Chairman of the Board

Copyright © 2011 The Hampton-Brown Company, Inc., a wholly owned subsidiary of the National Geographic Society, publishing under the imprints National Geographic School Publishing and Hampton-Brown.

All rights reserved. No part of this book may be reproduced or transmitted in any form or by any means, electronic or mechanical, including photocopying, recording, or by an information storage and retrieval system, without permission in writing from the Publisher.

National Geographic and the Yellow Border are registered trademarks of the National Geographic Society.

National Geographic School Publishing
Hampton-Brown
www.myNGconnect.com

Printed in the USA.
RR Donnelley
Jefferson City, MO

ISBN: 978-0-7362-7794-5

10 11 12 13 14 15 16 17 18 19 20

1 2 3 4 5 6 7 8 9 10

PROGRAM REVIEWERS

Carol Brueggeman
K–5 Science/Math Resource Teacher
District 11
Colorado Springs, CO

Miranda Carpenter
Teacher, MS Academy Leader
Imagine School
Bradenton, FL

Samuel Carpenter
Teacher
Coonley Elementary
Chicago, IL

Diane E. Comstock
Science Resource Teacher
Cheyenne Mountain School District
Colorado Springs, CO

Kelly Culbert
K–5 Science Lab Teacher
Princeton Elementary
Orange County, FL

Karri Dawes
K–5 Science Instructional Support Teacher
Garland Independent School District
Garland, TX

Richard Day
Science Curriculum Specialist
Union Public Schools
Tulsa, OK

Michele DeMuro
Teacher/Educational Consultant
Monroe, NY

Richard Ellenburg
Science Lab Teacher
Camelot Elementary
Orlando, FL

Beth Faulkner
Brevard Public Schools
Elementary Training Cadre,
Science Point of Contact,
Teacher, NBCT
Apollo Elementary
Titusville, FL

Kim Feltre
Science Supervisor
Hillsborough School District
Newark, NJ

Judy Fisher
Elementary Curriculum Coordinator
Virginia Beach Schools
Virginia Beach, VA

Anne Z. Fleming
Teacher
Coonley Elementary
Chicago, IL

Becky Gill, Ed.D.
Principal/Elementary Science Coordinator
Hough Street Elementary
Barrington, IL

Rebecca Gorinac
Elementary Curriculum Director
Port Huron Area Schools
Port Huron, MI

Anne Grall Reichel Ed. D.
Educational Leadership/ Curriculum and Instruction Consultant
Barrington, IL

Mary Haskins, Ph.D.
Professor of Biology
Rockhurst University
Kansas City, MO

PROGRAM REVIEWERS

Arlene Hayman
Teacher
Paradise Public School District
Las Vegas, NV

DeLene Hoffner
Science Specialist, Science
Methods Professor,
Regis University
Academy 20 School District
Colorado Springs, CO

Cindy Holman
District Science Resource
Teacher
Jefferson County Public Schools
Louisville, KY

Sarah E. Jesse
Instructional Specialist for
Hands-on Science
Rutherford County Schools
Murfreeboro, TN

Dianne Johnson
Science Curriculum Specialist
Buffalo City School District
Buffalo, NY

Kathleen Jordan
Teacher
Wolf Lake Elementary
Orlando, FL

Renee Kumiega
Teacher
Frontier Central School District
Hamburg, NY

Edel Maeder
K-12 Science Curriculum
Coordinator
Greece Central School District
North Greece, NY

Trish Meegan
Lead Teacher
Coonley Elementary
Chicago, IL

Donna Melpolder
Science Resource Teacher
Chatham County Schools
Chatham, NC

Melissa Mishovsky
Science Lab Teacher
Palmetto Elementary
Orlando, FL

Nancy Moore
Educational Consultant
Port Stanley, Ontario, Canada

Melissa Ray
Teacher
Tyler Run Elementary
Powell, OH

Shelley Reinacher
Science Coach
Auburndale Central Elementary
Auburndale, FL

Kevin J. Richard
Science Education Consultant,
Office of School Improvement
Michigan Department of
Education
Lansing, MI

Cathe Ritz
Teacher
Louis Agassiz Elementary
Cleveland, OH

Rose Sedely
Science Teacher
Eustis Heights Elementary
Eustis, FL

Robert Sotak, Ed.D.
Science Program Director,
Curriculum and Instruction
Everett Public Schools
Everett, WA

Karen Steele
Teacher
Salt Lake City School District
Salt Lake City, UT

Deborah S. Teuscher
Science Coach and
Planetarium Director
Metropolitan School District
of Pike Township
Indianapolis, IN

Michelle Thrift
Science Instructor
Durrance Elementary
Orlando, FL

Cathy Trent
Teacher
Ft. Myers Beach Elementary
Ft. Myers Beach, FL

Jennifer Turner
Teacher
PS 146
New York, NY

Flavia Valente
Teacher
Oak Hammock Elementary
Port St. Lucie, FL

Deborah Vannatter
District Coach, Science
Specialist
Evansville Vanderburgh School
Corporation
Evansville, IN

Katherine White
Science Coordinator
Milton Hershey School
Hershey, PA

Sandy Yellenberg
Science Coordinator
Santa Clara County Office
of Education
Santa Clara, CA

Hillary Zeune de Soto
Science Strategist
Lunt Elementary
Las Vegas, NV

LIFE SCIENCE
CONTENTS

What is Life Science?... 2
Meet a Scientist 4

CHAPTER 1

How Do Scientists Classify Living Things?..... 5
Science Vocabulary 8
Classifying Living Things 10
Classifying Plants 16
Classifying Invertebrates 20
Classifying Vertebrates 28
   NATIONAL GEOGRAPHIC Discovering a Species in Madagascar 38
Conclusion and Review 40
   NATIONAL GEOGRAPHIC **LIFE SCIENCE EXPERT:** Canopy Biologist 42
   NATIONAL GEOGRAPHIC **BECOME AN EXPERT:** Frightful Animals:
   Just Trying to Survive 44

CHAPTER 2

What Are the Interactions in Ecosystems? 53

Science Vocabulary 56

Ecosystems 58

Predation and Competition 62

Living Together 66

Changing Communities 70

Humans Change Ecosystems 76

NATIONAL GEOGRAPHIC Rooftop Gardens 84

Conclusion and Review 86

NATIONAL GEOGRAPHIC LIFE SCIENCE EXPERT: Conservationist 88

NATIONAL GEOGRAPHIC BECOME AN EXPERT: Interactions in a Coral Reef 90

CHAPTER 3

How Does Energy Move in an Ecosystem?....101

- Science Vocabulary 104
- **Producers and Consumers** 106
- **Herbivores, Carnivores, and Omnivores** 110
- **Decomposers** 114
- **Food Chains and Food Webs** 116
- NATIONAL GEOGRAPHIC Deep-Sea Vents: Living on the Edge 120
- Conclusion and Review 122
- NATIONAL GEOGRAPHIC **LIFE SCIENCE EXPERT:** Ecologist 124
- NATIONAL GEOGRAPHIC **BECOME AN EXPERT:** Energy in Ecosystems: The Arctic 126

CHAPTER 4

How Do Living Things Survive and Change?...133

- Science Vocabulary 136
- **Physical Characteristics of Living Things** 138
- **Behaviors Help Animals Survive** 144
- NATIONAL GEOGRAPHIC All Hands on Deck! Saving Right Whales 152
- **Life Cycle Adaptations** 156
- **When Environments Change** 160
- Conclusion and Review 168
- NATIONAL GEOGRAPHIC **LIFE SCIENCE EXPERT:** Aquatic Ecologist 170
- NATIONAL GEOGRAPHIC **BECOME AN EXPERT:** The Amazing World of Ants 172

TECHTREK
myNGconnect.com

 Student eEdition
 Vocabulary Games
 Digital Library
 Enrichment Activities

CHAPTER 5

How Do Body Systems Work Together?......181

Science Vocabulary 184

Organ Systems 186

Skeletal and Muscular Systems 188

Circulatory and Respiratory Systems 192

Digestive and Excretory Systems 198

The Nervous System 202

Comparing Organ Systems 206

 Making Sense of Senses 210

Conclusion and Review 214

LIFE SCIENCE EXPERT: Surgeon 216

BECOME AN EXPERT: Ironman Triathlon Race: Organ Systems Working Together......................... 218

Glossary..EM1
Index EM5

LIFE SCIENCE

What Is Life Science?

Life science is the study of all the living things around you and how they interact with one another and with the environment. This type of science investigates how living things are similar to and different from one another, how they live and reproduce, and how they function in the environment. Life science includes the study of humans, as well as all the other kinds of living things on Earth. People who study living things and the environment are called life scientists.

You will learn about these aspects of life science in this unit:

HOW DO SCIENTISTS CLASSIFY LIVING THINGS?

Scientists classify living things into groups based on their characteristics. The classification system begins with the largest group, a kingdom, and is divided into smaller groups. Life scientists use the classification system to study different living things.

WHAT ARE THE INTERACTIONS IN ECOSYSTEMS?

Ecosystems are filled with living and nonliving things. The living things and nonliving things interact with one another. Life scientists study interactions in ecosystems and learn what causes ecosystem changes.

HOW DOES ENERGY MOVE IN AN ECOSYSTEM?

Living things need energy. Some living things get energy from sunlight. Other living things get energy from the food they eat. Life scientists study how energy moves through an ecosystem.

HOW DO LIVING THINGS SURVIVE AND CHANGE?

Plants and animals show a great variety of adaptations that enable them to survive in their environments. If the environment changes, those living things that cannot survive move to new environments or die out. Life scientists study the adaptations of living things.

HOW DO BODY SYSTEMS WORK TOGETHER?

Humans have body systems that work together to carry out life processes. Other kinds of living things also have body systems that carry out similar tasks. Life scientists study the way the human body works and make comparisons to other living things.

NATIONAL GEOGRAPHIC

MEET A SCIENTIST

Maria Fadiman: Ethnobotanist

Maria Fadiman is an ethnobotanist and National Geographic Emerging Explorer. She was born with a passion for conservation and a fascination with indigenous cultures. Ethnobotany lets her bring it all together. Ethnobotany is the scientific study of the relationships that exist between people and plants.

On her first trip to the rain forest, Maria met a woman who was in terrible pain because no one in her village could remember which plant would cure her. Through this situation, Maria saw that knowledge of plant uses was being lost. In that moment she knew that learning more about indigenous people's use of plants was what she wanted to do with the rest of her life.

Of her work, Maria says, "I used to think that going to the jungle made my life an adventure. However, after years of unusual work in exotic places, I realize that it is not how far off I go, or how deep into the forest I walk that gives my life meaning. I see that living life fully is what makes life—anyone's life, no matter where they go or do not go—an adventure."

LIFE SCIENCE

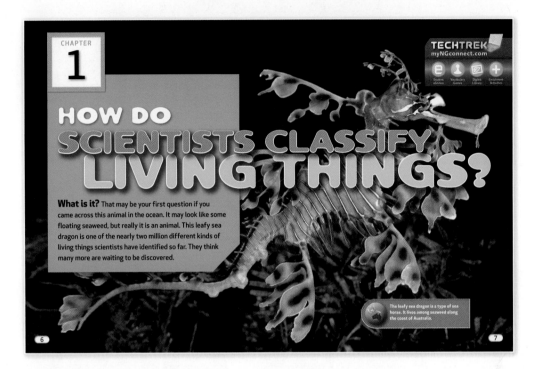

After reading Chapter 1, you will be able to:

- Explain how similarities are the basis of classification. **CLASSIFYING LIVING THINGS**
- Identify and explain the classification system of living things.
  CLASSIFYING LIVING THINGS
- Classify plants as vascular and nonvascular. **CLASSIFYING PLANTS**
- Classify animals as invertebrates and vertebrates. **CLASSIFYING INVERTEBRATES, CLASSIFYING VERTEBRATES**
- **Science in a Snap!** Classify animals as vertebrates and invertebrates.
  CLASSIFYING INVERTEBRATES

CHAPTER

1

HOW DO SCIENTISTS LIVING

What is it? That may be your first question if you came across this animal in the ocean. It may look like some floating seaweed, but really it is an animal. This leafy sea dragon is one of the nearly two million different kinds of living things scientists have identified so far. They think many more are waiting to be discovered.

CLASSIFY THINGS?

 The leafy sea dragon is a type of sea horse. It lives among seaweed along the coast of Australia.

SCIENCE VOCABULARY

cell (SEL)

A **cell** is the smallest unit of a living thing. (p. 11)

Plants are made up of cells.

kingdom (KING-dum)

A **kingdom** is one of the six main groups into which living things are sorted. (p. 12)

Mushrooms are classified in the fungi kingdom.

species (SPĒ-sēz)

A **species** is a group of similar living things that can produce offspring who can also produce offspring. (p. 14)

This species of camel has two humps.

my Science Vocabulary

cell (SEL)
invertebrate (in-VUR-tuh-brit)
kingdom (KING-dum)
species (SPĒ-sēz)
vascular plant (VAS-kyū-lar PLANT)
vertebrate (VUR-tuh-brit)

TECHTREK myNGconnect.com
Vocabulary Games

vascular plant
(VAS-kyū-lar PLANT)

A **vascular plant** is one that contains bundles of tubelike cells that transport water and food throughout the plant. (p. 17)

In these vascular plants, water is transported from the roots to the leaves and flowers.

invertebrate
(in-VUR-tuh-brit)

An **invertebrate** is an animal with no backbone. (p. 20)

A jelly is an invertebrate because it has no backbone.

vertebrate (VUR-tuh-brit)

A **vertebrate** is an animal with a backbone. (p. 28)

This tiny hummingbird is a vertebrate because it has a backbone.

Classifying Living Things

How are the animals you see in the photo alike and different? The animals have different colors, but they all have hair covering their bodies. These animals also give birth to live young, and the young feed on milk from their mothers' bodies. Because of these characteristics, these animals are classified as mammals.

Classification is useful in science. If an animal is classified as a bird, for example, then scientists all over the world know certain things about it. They know it is related to other birds and has certain things in common with them. For example, it breathes air and has feathers and wings.

These animals share certain characteristics that identify them as mammals.

Cells To classify living things, scientists look for likenesses between them. All living things are made of cells. A cell is the smallest unit of an organism. You are made of cells. So are a tree, a beetle, and a mushroom. Some living things have many more cells than others. Some have just one cell.

The cells of living things are different. If you look at a plant cell it has a cell wall. This gives the plant support so it can grow tall. Animal cells do not have a cell wall. Animals need some other way to support their bodies, such as bones.

The cells of living things have certain structures and functions. These differ from one living thing to another. Scientists use cells to help them classify living things.

SOME PARTS OF A PLANT CELL

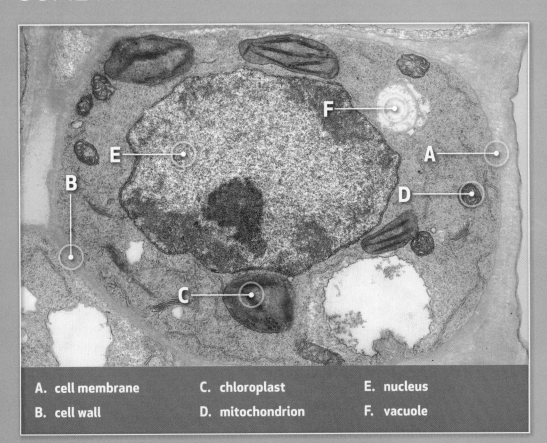

A. cell membrane
B. cell wall
C. chloroplast
D. mitochondrion
E. nucleus
F. vacuole

Kingdoms Scientists group similar types of living things together. A kingdom is one of the six main groups into which living things are sorted.

At one time, scientists thought all organisms were either plants or animals. So they classified every living thing into the plant kingdom or animal kingdom. Then scientists found organisms that did not fit into either of these kingdoms. As scientists learned more about organisms, they developed better ways to classify them. Today most scientists use a system with six kingdoms.

Look at the organisms in the photo on the right. Which kingdoms would you classify them in?

KINGDOM

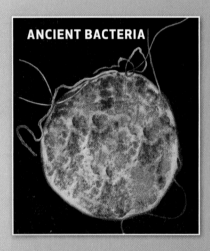

ANCIENT BACTERIA

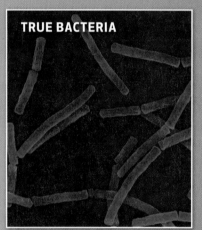

TRUE BACTERIA

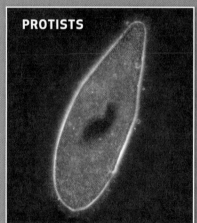

PROTISTS

CHARACTERISTICS

- only one cell
- make their own food
- often live in harsh places, such as hot springs or deep underground

- only one cell
- some make their own food; some must get their food
- live on land or in water

- usually one cell
- plant-like protists make food; animal-like or fungi-like protists get food
- live in water or wet places

CLASSIFICATION OF A BACTRIAN CAMEL

Observe that the animals in each group become more similar.

KINGDOM: ANIMAL

PHYLUM: CHORDATA

CLASS: MAMMALIA

ORDER: ARTIODACTYLA

FAMILY: CAMELIDAE

GENUS: *CAMELUS*

SPECIES: *CAMELUS BACTRIANUS*

Before You Move On

1. How is classification useful to scientists?
2. What is a cell?
3. **Draw Conclusions** You see a furry animal in a tree. Could the animal be a bird? Explain your conclusion.

Classifying Plants

Observe the plants on this page. How are the trees, the ferns, and the mosses alike? How are they different? All three are kinds of plants. That means they have many cells and make their own food. They have structures in their cells that use light energy to make food from water and carbon dioxide. Also, plants don't move from place to place as animals do.

But plants have differences, too. For example the trees in the picture are very tall. The moss, in comparison, grows low to the ground. A plant's size, shape, and where it grows are all differences.

Vascular and Nonvascular

Trees and mosses have many different characteristics. One of the most important is that trees are vascular plants, and mosses are not. A vascular plant contains cells that form a system of tubes throughout the plant. The tubes carry water and food from one part of the plant to another.

The tubes also help support the plant. Vascular plants can vary in height. Some are small but others, like trees, can grow very tall.

Mosses, on the other hand, cannot grow tall. They are nonvascular plants. Nonvascular plants do not have a system of tubes to carry water and food throughout the plant. Mosses seldom grow taller than 15 centimeters (6 inches).

fern

These tiny liverworts are nonvascular plants.

Reproduction Plants can also be grouped by how they reproduce, or make new plants. Compare the three groups of vascular plants.

Flowering Plants All of the plants in this photo have one thing in common. Flowers are the way they reproduce. Most plants on Earth are flowering plants. An oak tree is a flowering plant, and the grass in the park is, too. Flowers are not always big and showy but they all produce seeds. Seeds contain tiny undeveloped plants and a supply of food for the plant. The seeds are covered by a fruit. Just like flowers, there are many types of fruits.

You are familiar with many fruits like peaches and tomatoes. But even daffodils and poppies produce seed covers called fruits. A fruit is simply a covering that protects the seed until conditions are right for the seed to grow.

Flowers, such as California poppies and yellow goldfields, grow wild in this field.

Conifers Trees that grow cones are called conifers. Pines and other trees that have cones do not have flowers. But they still have seeds. The seeds grow on the scales that make up the seed cone. Conifer seeds are not covered by fruits.

Ferns Plants called ferns make spores to reproduce. Spores are like seeds but they do not contain stored food. They develop in spore cases on the underside of a fern's leaves. When the time is right, the spore cases burst open and launch the tiny spores into the air.

The scales of this cone have opened up to release the seeds.

These black dots are spore cases on the underside of fern leaves.

Before You Move On

1. Describe a vascular plant.
2. How are a California poppy and a pine tree alike and different in how they reproduce?
3. **Apply** Would a garden center sell fern seeds? Explain your answer.

Classifying Invertebrates

Scientists put all animals into two large groups—animals with backbones and animals without backbones. You may think a backbone is pretty common. After all, you have a backbone. However, less than 5 percent of all known animal species have backbones. Animals without a backbone are called invertebrates.

There are several different groups of invertebrates and they can be found almost everywhere on Earth.

Sponges Sponges are very simple animals with no stomachs or other organs. Sponges do have special cells that keep water flowing through them. Water enters through holes in the sponge. The water contains food and oxygen. The sponge's cells take in the food and oxygen, and the water leaves through a larger opening.

People used to think sponges, such as red finger sponges, were plants.

FUNGI

PLANTS

ANIMALS

- most have many cells
- get food from plants or animals
- live on land

- all have many cells
- use sunlight to make their own food
- most live on land; some live in water

- all have many cells
- eat plants or other animals
- live on land or in water

Classification System

Think about what you do when you look up the definition of a word in the dictionary. You start by turning to the section of the dictionary that includes the first letter of the word. Then you search for the second letter, then the third letter and so on, until you find the word. Each group of letters includes fewer and fewer words. Also, each group includes words that are more and more similar in how they are spelled.

Scientists classify living things in a similar way. They start by deciding into which kingdom the organism belongs. Then the organism is placed into smaller and smaller groups based on similarities with other members of that group. The classification groups, or levels, within a kingdom are phylum, class, order, family, genus, and species. As you move from kingdom to species, the living things in that group are more and more similar. A **species** is a group of similar living things that can mate and produce offspring that can also mate and produce offspring.

Each species has a two-word scientific name. The first name is the genus; the second name is the species. The name for this camel is *Camelus bactrianus*.

Jellies, Anemones, and Corals

If you have ever visited an ocean beach, you may have been warned to watch out for jellyfish, or jellies. That's because they sting. A jelly has no bones, heart, or brain, but its cells are organized into tissues. Tentacles hang down around its mouth on the underside of its body. The tentacles sting small ocean animals.

An anemone is like an upside-down jelly. Its tentacles and mouth face upward. An anemone can creep along. But usually this invertebrate just waits on a rock for a meal to come along.

Corals also have stinging tentacles for getting food. Thousands of small coral animals live together in colonies. They each produce a hard skeleton around their soft bodies.

Moon jellies grow to about 38 centimeters (15 inches) wide and can be found in every ocean.

This frilled sea anemone has its tentacles extended for feeding.

Worms

You may have seen worms on the sidewalk after it rains, or in the soil while digging in a garden. These are earthworms. Many other kinds of worms exist. Worms can be flat, round, or made up of small sections. Some are very colorful. Some live on land, while others live in water. Some even live inside other animals.

Worms are more complex than jellies, anemones, and coral. Worms have tissues that form organs and organ systems. Earthworms, for example, have organs, such as muscles and blood vessels. They do not have eyes, but special cells allow them to tell dark from light.

Earthworms live in moist places such as soil or leaf litter.

This colorful flatworm lives in the ocean.

Snails and Octopuses

It may surprise you that snails and octopuses are closely related. They are both soft-bodied invertebrates. They are classified together because they both have a muscular foot. The snail glides along on its foot. The rest of the snail is covered with a hard shell. The octopus's foot is split to form its tentacles. The octopus has a shell, too, but it is inside its body.

A snail can pull itself into its shell for protection.

The suckers on an octopus's eight arms can feel, smell, and taste.

Sea Stars Another kind of invertebrate is the sea star. Some people call them star fish but they are not fish. A sea star has a spiny skin. Its arms radiate out from its center. Sea stars have structures like suction cups underneath their arms that they use to hold on to rocks and to break open the shells of shellfish so they can eat them. To eat, the sea star pushes its stomach right out of its mouth and into its prey. It digests its food and then pulls its stomach back inside.

These red sea stars live in the Pacific Ocean near the Galápagos Islands.

Arthropods Arthropods are invertebrates that have legs with joints and a body divided into sections. They also have an exoskeleton, which is a hard outer covering that protects the animal's soft body parts inside. There are about one million different kinds of arthropods on Earth. That is more than any other kind of animal.

Arachnids Both spiders and scorpions are arachnids. Many people think spiders are insects, but they're not. Spiders have two body parts, eight legs, and up to eight eyes. Spiders have fangs but they do not have teeth. They eat by injecting digestive juices into their prey and sucking out the food.

Crustaceans This group includes crabs, lobsters, and shrimp. They have five pairs of legs, some of which may form claws. All crustaceans breathe through gills and most live in water.

Count the legs on this tarantula.

Count the legs on this Sally Lightfoot crab.

Insects The largest group of invertebrates is insects. All insects have six legs and three main body parts. Like all arthropods, insects have an exoskeleton. An exoskeleton does not grow with the animal. Insects, like other arthropods, shed their outer covering when they grow. They then grow a larger exoskeleton. Insects have other body parts in common. Most have wings, antennae, and compound eyes or eyes that are made up of several smaller eyes.

Scientists can classify insects into two groups based on the stages of their life cycles. Some insects, like grasshoppers and cockroaches, have three stages in their life cycles. The young looks like the adult and it sheds its exoskeleton as it grows bigger. Other insects, such as bees and butterflies, have a four-stage life cycle. The young of these insects looks very different from the adult.

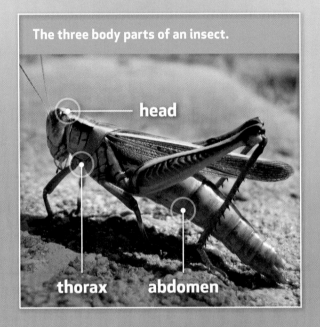

The three body parts of an insect.

Science in a Snap! Using a Dichotomous Key

Observe the photos. Read each pair of statements. Decide if each statement describes one or more insects. If only one insect from the group matches the description, the insect will be identified.

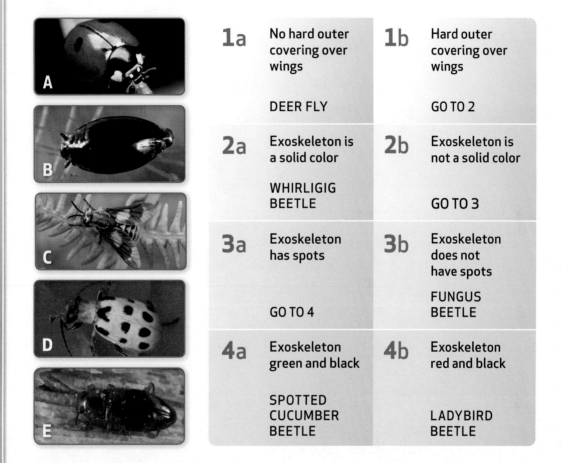

1a	No hard outer covering over wings DEER FLY	**1b**	Hard outer covering over wings GO TO 2
2a	Exoskeleton is a solid color WHIRLIGIG BEETLE	**2b**	Exoskeleton is not a solid color GO TO 3
3a	Exoskeleton has spots GO TO 4	**3b**	Exoskeleton does not have spots FUNGUS BEETLE
4a	Exoskeleton green and black SPOTTED CUCUMBER BEETLE	**4b**	Exoskeleton red and black LADYBIRD BEETLE

Have a partner select 5 tools for writing, such as pens and pencils. Develop a key like this one to identify each writing tool. What characteristic separates one tool from the rest for your first pair of statements?

Before You Move On

1. Why is a sea star classified as an invertebrate?
2. In what way are anemones like upside-down jellies?
3. **Evaluate** Which do you think is a more complex invertebrate, a sponge or an earthworm? Explain.

Classifying Vertebrates

What do horses and mice have in common? Both of them have backbones. These animals are called vertebrates. The backbone of all vertebrates grows as the animal grows. The backbone surrounds and protects the spinal cord in all vertebrates. The backbone bends, allowing the animal to do complex movements.

Most known species of fish live in salty ocean water, but some live in freshwater lakes and rivers.

Fishes All fishes are vertebrates because they all have backbones. Goldfish, trout, and many other fishes have skeletons made of bone. But some fishes, including sharks, have cartilage instead of bone. Cartilage is softer than bone but it is flexible and strong. The tip of your nose is made of cartilage.

The outside of a fish's body is usually covered with scales. Some fish have scales that feel smooth, while other fish have rough scales. The scales of a shark are shaped like tiny teeth. They feel like sandpaper.

Fishes do not have lungs. To breathe, a fish uses gills. Water comes in through the fish's mouth and passes over its gills. The surfaces in the gills absorb, or take in, oxygen from the water. Then the water passes out through gill openings on the sides of the fish's head.

TECHTREK
myNGconnect.com

All fishes live only in water. Their body parts are adapted for swimming.

Digital Library

fin

gill opening

scales

Amphibians Amphibians include frogs, toads, newts and salamanders, and caecilians. Amphibians spend part of their lives in water and part on land. All amphibians need a moist environment in order to survive. However, not all amphibians are the same. Some, like frogs and toads, have powerful back legs that allow them to jump or hop. Others, like newts and salamanders, have four legs and a tail. Caecilians have no legs at all. Toads may have thick, bumpy skin, but amphibians generally have smooth skin. They do not have scales.

Observe the bumpy skin on this Couch's spadefoot toad.

Salamanders hatch from eggs that are often laid in ponds or moist soil.

Amphibians are the only vertebrates to go through metamorphosis. This means they make a complete change from one form to another as they develop.

Frogs, for example, lay their eggs in water. The tadpoles that hatch from the eggs live in the water. Like fish, tadpoles use gills to breathe. Then metamorphosis occurs, and the tadpoles begin to change. They grow legs, lose their tails, and develop lungs. They become adults and live on land.

Amphibian tadpoles breathe through gills.

Adult amphibians breathe using lungs.

TECHTREK
myNGconnect.com

Digital Library

Reptiles

The pictures show two kinds of reptiles, a lizard and a snake. What can you learn about these animals by observing them?

You probably noticed that both animals are covered with scales. Many lizards and snakes live in hot, dry places. Their overlapping scales help trap moisture.

Most lizards have four legs and toes with claws. Snakes have no legs, but they use their strong muscles and flexible backbone to slither along the ground.

This species of lizard is called a semaphore gecko. The color of the gecko's scales helps it blend with its surroundings.

- leg
- scales
- clawed toe

Crocodiles, alligators, and turtles are reptiles too. Like all reptiles, these animals use lungs to breathe. If they go underwater, they must come up to breathe. Sometimes they keep just their nose above the surface so they can breathe.

All reptiles behave in certain ways to help them control their body temperature. When it is cool out reptiles bask in the sun to warm their bodies. When it is hot they move to the shade. Their body temperature affects how quickly they can move.

Like all snakes, this green tree python does not have eyelids. It has scales that protect its eyes.

scales

Birds Have you ever been told you "eat like a bird"? Then you probably weren't eating much. However, birds actually eat a lot! Some even eat twice their weight each day. Birds need lots of energy to fly. All birds have feathers, two legs covered with scales, and wings.

Most birds have lightweight, hollow bones. Lighter bones make it easier for birds to fly. Birds hatch from eggs and use lungs to breathe air. Birds' feathers help them fly and also keep them warm even in the coldest environment.

A hummingbird beats its wings more than fifty times per second. That's fast enough to make a humming sound.

wing

feathers

beak

Hummingbirds are the smallest birds on Earth. Look at the broadtailed hummingbird. It has a long, strawlike beak that lets it reach far into a flower and suck out the nectar. Hummingbirds can move their wings farther to the back and front than most birds. This allows them to hover or stay in place while they feed. They move their wings so fast that their wings look like a blur.

Ostriches are the largest birds on Earth. They are taller than an adult human. An ostrich has wings, but the wings are small. In fact, an ostrich is too big to fly, but it sure can run! It can run 70 kilometers (42 miles) per hour on its strong, long legs.

Ostriches have powerful legs. They can run from danger or kick to defend themselves.

Mammals Cats, dogs, hamsters, and cows—these familiar animals are mammals. So are humans. But some mammals aren't so well known. Observe the narwhal and tamarin in the pictures. How are these animals like cats and dogs? What makes a mammal a mammal?

All mammals have fur or hair. The bodies of mammals produce heat, and fur traps that heat. This helps the mammal keep its body temperature about the same.

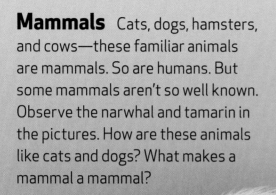

The golden lion tamarin is a small monkey that lives in the coastal forests of Brazil.

Most mammals have four limbs, which can be legs, arms, or flippers. Whales are an exception. They have flippers in front, but no back limbs. Mammals use lungs to breathe. Ocean mammals, like the narwhal, must come to the surface to breathe.

Most mammals give birth to live young, rather than hatching their young from eggs. One or both parents take care of the young. The female produces milk to feed them.

Like all vertebrates, mammals have backbones. Mammals also have skulls to protect the brain, and most mammals have teeth. Narwhals are known for having one amazing, very long tooth!

A narwhal does not have much fur or hair—only a few bristles—but enough to help classify it as a mammal.

Before You Move On

1. What body covering makes a bird different from other vertebrates?
2. Compare an amphibian and a reptile.
3. **Classify** A rabbit has fur and gives birth to live young. How would you classify this animal?

NATIONAL GEOGRAPHIC

DISCOVERING A SPECIES IN MADAGASCAR

Madagascar is a large island off Africa's southeast coast. Many animals and plants are found nowhere else in the world except Madagascar and neighboring islands. That includes lemurs. Theses animals are part of a small group of mammals known as primates. Humans, gorillas, and chimpanzees are primates, too.

Mireya Mayor never dreamed she would one day be tracking down lemurs in the forests of Madagascar. After all, she was a city girl from the United States. But lemurs fascinated Mayor. So when she found that little was known about them, she decided to do something about it. She headed for Madagascar.

Mireya Mayor looking for lemurs in the forests of Madagascar.

Earth's largest living lemur is the size of a house cat, but some lemurs are as small as mice. And that is what Mayor discovered in Madagascar—a tiny species of mouse lemur that had never been seen before by scientists.

This mouse lemur may be the smallest primate on Earth. It has ten fingers, ten toes, and a long furry tail. At night, the little lemur scurries around in trees to find food to eat.

The honor of naming a newly discovered species goes to the discoverer. Mayor named this lemur species *Microcebus mittermeieri*, or Mittermeier's Mouse Lemur. She named it after Russell Mittermeier, the man who encouraged Mayor to study lemurs.

This species of lemur is called a sifaka.

Mireya Mayor holds the mouse lemur she discovered in Madagascar.

Scientists classify living things into groups based on their shared characteristics. Living things are organized into six kingdoms—ancient bacteria, true bacteria, protests, fungi, plants, and animals. Each kingdom contains similar living things, but there are differences among the members of a kingdom, too. For example, not all plants have seeds, and not all animals have backbones.

The kingdoms are divided into smaller and smaller groups. A species is the smallest group. As you move from kingdom to species, the animals in the group are more and more similar.

Big Idea Scientists classify living things into groups based on similarities.

Vocabulary Review

Match each of the following terms with the correct definition.

A. kingdom
B. vertebrate
C. invertebrate
D. cell
E. vascular plant
F. species

1. An animal with no backbone
2. One of the main groups into which living things are sorted
3. The smallest unit of a living thing
4. A group of similar living things that can mate and produce offspring that can also produce offspring
5. An animal with a backbone
6. A plant with bundles of tubes that transport water and food around the plant

40

Big Idea Review

1. **Recall** What do scientists call a group of similar living things that can produce offspring that can also produce offspring?

2. **List** What are some characteristics of fish? List at least three.

3. **Contrast** How are fungi and plants different in the way they get their food?

4. **Classify** Make a chart to classify these animals as vertebrates and invertebrates. List these animals in the correct columns: bird, sponge, worm, spider, fish, octopus, snail, anemone, camel, human.

5. **Apply** What things would you look for in order to tell if an animal you find is an insect?

6. **Generalize** Would you be likely to find a new species of snake near the North Pole? Use what you know about reptiles to tell why or why not.

Write About Classification

Explain Write a paragraph out the animal in the picture below. Compare its characteristics with those of other animals.

spotted salamander

NATIONAL GEOGRAPHIC

CHAPTER 1 — LIFE SCIENCE EXPERT: CANOPY BIOLOGIST

What's It Like to Explore the Treetops? Ask a Canopy Biologist.

Bugs, leaves, and rope walkways—they're all part of the job for Meg Lowman. She works in the canopy, among the uppermost rain forest branches.

What do you do as a canopy biologist?

I act like a detective in the world's treetops. I discover new species. Then I try to figure out how everything works together in the forest as one big system to keep the entire planet healthy. I specialize in insects that eat plants. A lot of my time is spent observing beetles feeding on leaves. Beetles seem to be the biggest group of leaf-munchers! I try to learn which bug lives where, what it eats, and how it affects the health of trees.

When does your workday start and finish?

My day starts at sunrise. But I often work into the night. Many rain forest insects feed at night to avoid being eaten by birds. Sometimes you can actually hear the activity of insects feeding in the treetops!

What part of your job is most important to you?

My most important work has been to save rain forests, one at a time, by building canopy walkways. The walkways attract tourists. This allows local people to get income through tourism, not through cutting down trees to make logs. Let's face it; forests are worth more alive than dead!

Meg Lowman is amazed by the thousands of creatures she finds high in the rain forest canopy.

How could someone prepare to be a canopy biologist?

Spend time outdoors, read about nature, and volunteer at a local nature center or park. When I was growing up, my friends and I built tree forts. And we tried to heal earthworms that got injured by lawnmowers. Playing outdoors and observing nature is important for developing skills as a biologist. Also, take as much science as your school offers. Then study science and biology in college.

This canopy walkway is located in the rain forests of Belize, a country in Central America.

Lowman walks across a canopy raft to collect insect samples with her net.

NATIONAL GEOGRAPHIC

BECOME AN EXPERT

Frightful Animals: Just Trying to Survive

You've probably seen lots of pictures of members of the animal **kingdom**. Did any of them make you shiver with fright? Maybe you are not a fan of creatures that creep and crawl. Maybe you wouldn't want to be too close to an animal with a mouthful of razor-sharp teeth.

Many frightful animals are deadly, but usually not to people. The **species** on the following pages are among the deadliest on Earth. Think about how each one uses its body parts to get food and survive.

> Gila monsters are colorful lizards. They can be black with orange, yellow, or pink markings.

kingdom
A **kingdom** is one of the main groups into which living things are sorted.

species
A **species** is a group of similar living things that can mate and produce offspring that can also produce offspring.

Desert Monster

In the deserts of southwestern United States and western Mexico, you can find one of the world's most dangerous lizards—the Gila monster. This species of reptile isn't your everyday lizard. Once its powerful jaws crush down on something, it doesn't let go. When the lizard catches an animal, it bites hard. Poisonous venom is injected into the prey to kill it. The Gila monster usually eats small rodents, such as desert rats. It rarely attacks humans. Still, if you happen to see a Gila monster, you'd want to walk the other way.

range of the Gila monster

A Gila monster delivers venom through grooves in its teeth.

45

BECOME AN EXPERT

Super-sized Spider

Deep in the rainforests of South America, you'll find the world's largest spider. It's the goliath tarantula. Like all spiders, the goliath tarantula is an **invertebrate**. But it doesn't need a backbone to overpower the frogs, beetles, and lizards it eats. The goliath tarantula sneaks up on these animals. Then it pounces! The spider lands on its victim and injects venom through its fangs.

The tarantula does not have teeth for chewing food. Instead, it spits a liquid onto its victim. The liquid breaks down the unlucky animal's body. Then the spider slurps up its meal.

range of the goliath tarantula

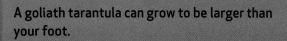

A goliath tarantula can grow to be larger than your foot.

invertebrate
An **invertebrate** is an animal with no backbone.

Night Stalker

You may not think of an owl as frightening—unless you're a small mammal running from it. The largest owl on Earth is the European eagle owl. Like all owls, the eagle owl is an expert hunter. It hunts at night, flying swiftly and silently through the darkness. It prefers small mammals, such as rabbits, squirrels, and chipmunks. However, its sharp beak and claws can take down a small deer!

range of the European eagle owl

The European eagle owl has a wingspan of 2 meters (6 feet).

BECOME AN EXPERT

Grisly Giggler

The spotted hyena is known for its giggle. Don't let that fool you. There's nothing funny about this dangerous killer.

This **vertebrate** lives on the grasslands of Africa. It lives in a group called a clan. A hyena clan is a noisy bunch. Its members communicate with cries, screams, and giggles.

Three things make the spotted hyena a super dangerous killer. First, it's fast. It can run 48 kilometers (30 miles) per hour and chase down almost any animal. Second, it usually hunts in a pack. Third, the hyena's strong jaws and teeth allow it to eat every bit of its victim, even bones.

range of the spotted hyena

Spotted hyenas hunt in packs and can bring down a water buffalo.

vertebrate
A **vertebrate** is an animal with a backbone.

Fierce Fish

Our next frightful animal doesn't giggle. It slices through the water of the world's oceans—silent and deadly. It's the great white shark. This fearsome fish grows to 6 meters (20 feet) long.

The open jaws of a great white are a fearful sight. Several rows of pointed teeth line its huge mouth. The teeth slant inward so that a victim in its jaws cannot escape.

Great whites generally don't attack people. They feed mostly on seals, sea lions, and sea turtles. But a shark swimming underwater may mistake a person on a surfboard for one of these animals and attack.

range of the great white shark

A great white shark has about 300 sharp teeth.

BECOME AN EXPERT

Fatal Hug

Here's a frightful animal that also has a mouthful of teeth. But it doesn't use them to kill. The blood python of Southeast Asia squeezes its victims to death.

The blood python lives mostly among the grasses and other vascular plants on the riverbanks and swamps of the rain forest. The snake lies in wait, sometimes for days, until a mammal comes along. Then it attacks. The python curls its strong, thick body around the victim and squeezes. Each time the victim tries to take a breath, the snake squeezes tighter. Eventually the doomed animal can no longer breathe. Then the snake swallows the animal whole.

range of the blood python

A blood python can grow to two meters (six feet) long. That's short and stubby for a python.

vascular plant
A **vascular** plant is one with bundles of tubelike cells that transport water and food around the plant.

Deadly Rings

If you happen to see a creature like the one in the photo along the rocky shores of Japan or Australia, here's some advice—stay away! It's a blue-ringed octopus, one of the world's deadliest animals. This octopus bites and injects venom into its victims. The venom is created by bacteria—living things made of single **cells** living in the octopus. The venom stops crabs and shrimp from moving. Then the octopus tears off pieces and sucks the flesh out of the shell or hard skin.

This deadly octopus may seem cruel, but it's not. Neither is the great white shark, blood python, or other "frightful" animals. They are simply trying to survive.

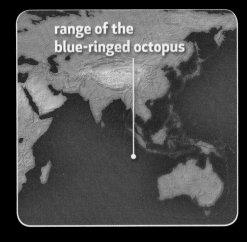

range of the blue-ringed octopus

The blue and black rings of the blue-ringed octopus become visible when the animal is frightened or disturbed.

cell
A **cell** is the smallest unit of a living thing.

BECOME AN EXPERT

CHAPTER 1: SHARE AND COMPARE

Turn and Talk What characteristics can help different kinds of animals survive? Form a complete answer to this question together with a partner.

Read Select two pages in this section that are the most interesting to you. Practice reading the pages so you can read them smoothly. Then read them aloud to a partner or small group. Talk about why the pages are interesting.

Write Write a conclusion that tells the important ideas about what you have learned about frightful animals. State what you think is the Big Idea of this section. Share what you wrote with a classmate. Compare your conclusions. Did your classmate recall which animals use venom?

Draw Think of another animal that seems particularly frightful. Draw a picture of it. Label features that make it frightful to you and to the animals it attacks. Include a caption that tells something interesting about the animal. Post your drawing with those of your classmates to create a Frightful Animals Hall of Fame.

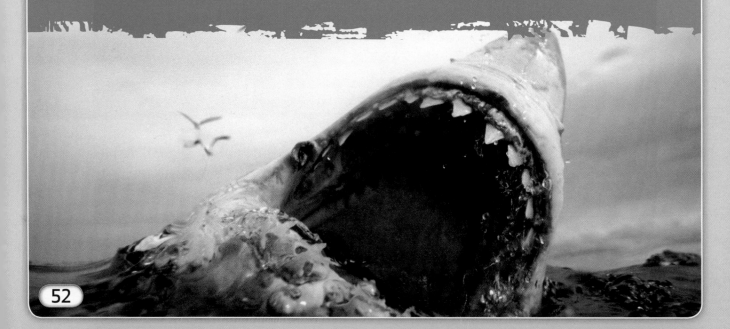

LIFE SCIENCE

After reading Chapter 2, you will be able to:

- Identify living and nonliving parts of ecosystems. **ECOSYSTEMS**
- Explain how ecosystems allow living things to survive. **ECOSYSTEMS**
- Contrast the role of predators and prey in an ecosystem. **PREDATION AND COMPETITION**
- Explain how competition for resources affects species in an ecosystem.
  PREDATION AND COMPETITION
- Describe three different kinds of symbiotic relationships between species.
  LIVING TOGETHER
- Describe natural processes that cause ecosystems to change over time.
  CHANGING COMMUNITIES
- Explain how invasive species can affect the balance of an ecosystem.
  CHANGING COMMUNITIES
- Explain how human activities affect ecosystems. **HUMANS CHANGE ECOSYSTEMS**
- Science in a Snap! Identify ways that humans can help the environment.
  HUMANS CHANGE ECOSYSTEMS

CHAPTER 2

WHAT ARE THE INTERACTIONS IN

This butterfly is sipping sweet nectar from the flower. As it feeds, the butterfly carries pollen from flower to flower. This pollinates the flowers. The interaction between the butterfly and flowers is helpful to both species. Their interaction is just one of the countless interactions taking place in their ecosystem every day.

The yellow flowers of this *Inula* attract many kinds of butterflies and bees.

TIONS ECOSYSTEMS?

SCIENCE VOCABULARY

ecosystem (Ē-kō-sis-tum)

An **ecosystem** is all the living and nonliving things in an area and their interactions. (p. 58)

The elephants, lions, and water hole are parts of this desert ecosystem.

population (pop-yū-LĀ-shun)

A **population** is all the members of a species living in a certain area. (p. 60)

These individuals are part of the meerkat population.

community (kuh-MYŪ-nuh-tē)

A **community** is all the different organisms that live and interact in an area. (p. 61)

The first community to grow after a fire includes fireweed and the animals that eat it.

my Science Vocabulary

community (kuh-MYŪ-nuh-tē)
ecosystem (Ē-kō-sis-tum)
niche (nēsh)
population (pop-yū-LĀ-shun)
succession (suhk-SESH-un)
symbiosis (sim-bē-Ō-sis)

TECHTREK
myNGconnect.com
Vocabulary Games

niche (nēsh)

A **niche** is the way an organism interacts with living and nonliving parts of its ecosystem. (p. 62)

The niche of the arctic fox includes eating lemmings and other small animals.

symbiosis (sim-bē-Ō-sis)

Symbiosis is the close association of members of different species that live together. (p. 66)

The symbiosis of acacia ants and bullhorn acacias benefits both species.

succession (suhk-SESH-un)

Succession is the gradual replacement of one community by another over time. (p. 71)

After many years, succession will change this beaver pond into a marsh.

Ecosystems

A water hole in the Kalahari Desert attracts many visitors. A herd of elephants comes to drink. Lions come to watch for prey. You might also see grazing animals, such as antelope and wildebeests. These are just a few of the many kinds of mammals that live here. The Kalahari is also home to many birds and reptiles, and countless insects.

Like all deserts, the Kalahari is very dry. You don't need to look far to find sand and rocks. In the dry season, it may not rain for months.

To you, the Kalahari may seem to be a harsh land. Yet the many organisms living here are able to find what they need to survive.

When the rains come, fields of grasses cover the land. These grasses use the bright sunlight to grow. Antelopes and wildebeests graze on the grasses. Lions and hyenas feed on the grazers. When not hunting, a lion might rest in the shade of a tree. Ground squirrels find shelter by digging burrows in the sandy soil.

The Kalahari Desert in southern Africa is home to elephants, lions, and acacia trees.

All of the organisms in the Kalahari interact with each other and their environment to form an ecosystem. The Kalahari ecosystem includes living plants, animals, fungi, and microscopic organisms. It also includes nonliving things, such as sunlight, water, air, and soil.

Earth has many ecosystems. A forest and a prairie are ecosystems. So are a pond and a stream. Organisms can only survive in ecosystems that meet their needs. For example, a rabbit can live in a prairie, but not in a pond. A fish can live in a pond, but not a prairie.

Populations and Communities

If you visited the Kalahari, you might see a meerkat. Look around. Chances are, others like it are nearby. Each meerkat is part of a **population**. A population is all the individuals of a species that live in an area. The area can be large or small. A population could be all the meerkats in the Kalahari or just those in one valley.

INDIVIDUAL One organism, such as a meerkat, is the smallest unit of an ecosystem.

POPULATION All the meerkats that live in one place are a population.

Populations of many different species live in the Kalahari desert. All the different organisms that live and interact in an area make up a **community**. The picture below shows just a few of the species in the Kalahari community. The community also includes all the insects, worms, fungi, and bacteria.

Members of the community are constantly interacting with nonliving factors in the environment. A meerkat breathes in air. A lizard suns itself on a rock. A tree takes in water and nutrients from the soil. All the members of the community and the nonliving parts of the environment with which they interact make up an ecosystem.

COMMUNITY All the populations of living things that live and interact in an area form a community.

ECOSYSTEM The ecosystem includes all the living and nonliving things that interact with each other in the Kalahari.

Before You Move On

1. What is an ecosystem?
2. List these words in order from the smallest unit to the largest unit: community, ecosystem, individual, population.
3. **Evaluate** Are all the fish in a lake a population? Why or why not?

Predation and Competition

Predation An arctic fox slowly steps through the summer grass. It has something in its sights. It is hunting a lemming—a small animal that looks like a mouse. The lemming is nibbling on the grass, unaware of the danger nearby. Suddenly the fox pounces. It snaps up the lemming in its powerful jaws. Now it's the fox's turn to eat.

It may seem cruel for the fox to hunt and kill the lemming. But the fox is only doing what it needs to survive.

The way that an organism interacts with the living and nonliving parts of its environment is called a niche. A niche is the role an organism plays in its ecosystem. You can think of an organism's niche as the way it makes its living. Part of the fox's niche is to kill and eat small animals.

This interaction in which one organism eats another is called predation. The organism doing the eating is the predator. The organism being eaten is the prey.

The arctic fox, the lemming, and the grass are all part of the tundra ecosystem in Alaska.

Different species eat different organisms. Bullfrogs, for instance, eat other animals, mostly insects. Some animals, such as the arctic fox, eat both animals and plants.

Predation causes ecosystems to stay balanced. One way it does this is by keeping prey populations from getting too large. When populations get too large, food becomes scarce. Many in the population die from lack of food and from disease. Predation helps control these factors.

Predation also weeds out the weaker individuals in a population. After all, it's easier for a fox to catch a slow, weak lemming than a strong one that can scurry away quickly. When weak individuals die, the population becomes stronger and healthier.

A desert locust makes a meal of a leaf.

Competition Would you build your house on the side of a cliff? Probably not, but puffins do. Puffins are sea birds. They spend most of their time over the ocean diving for fish, shrimp, and mollusks. They nest in cliffs along the ocean. Each spring, great numbers of puffins return to the cliffs to mate and lay eggs.

The cliffs keep the puffins safe from predators. With so many birds and so little space, puffins compete with each other for suitable sites. They also compete with rabbits and other kinds of birds. Only puffins that get good nesting sites are successful in raising their young.

The Atlantic puffin builds its nest in a burrow on the side of a cliff.

The cliffs along the coast of Iceland are a favorite nesting site for puffins.

Competition is an interaction in which organisms struggle to get the resources they need. Puffins compete for shelter on cliffs. Animals must compete for any resource that is in short supply. Often they compete for food and water. Usually the fastest or strongest individuals win the competition. Those that do not get enough resources may die or move away.

Plants also compete for resources, such as light, water, minerals, and living space. Trees in a forest grow tall to reach sunlight. Taller trees often shade out shorter trees. Plants that spread their roots wide and deep can get more water and minerals from the soil. Dandelions and other weeds compete with grass this way.

Atlantic puffins compete for space to raise their young.

Before You Move On

1. What is a niche?
2. How does predation help an ecosystem stay balanced?
3. **Draw Conclusions** Why do organisms have to compete for resources?

Living Together

The birds on the warthog may look like they're just catching a free ride. But in fact, the warthog and the yellow-billed oxpeckers on its back are helping each other survive. The birds eat ticks and insects that live on the warthog. That's good for the birds and the warthog. The birds get food. The warthog gets rid of ticks and insects, which drink its blood and can carry diseases.

The interactions between the oxpeckers and the warthog form a close association called **symbiosis**. In symbiosis, two species live together in a way that benefits, or helps, at least one of the species. In the case of the oxpeckers and warthog, both species benefit.

Yellow-billed oxpeckers live with warthogs in a way that benefits both species.

In some kinds of symbiosis, the species depend on each other so much that one could not survive without the other. That's the way it is with acacia ants and bullhorn acacia trees.

The large thorns of the bullhorn acacias serve as nesting sites for the acacia ants. The trees also produce nectar and yellow sacs of protein-rich food for the ants to eat. In return, these stinging ants drive away insects and larger animals that would eat the acacia leaves. The ants also prune vines and seedlings that threaten to grow over the trees.

This acacia ant is collecting the yellow, protein-rich food that grows on the leaves of the bullhorn acacia.

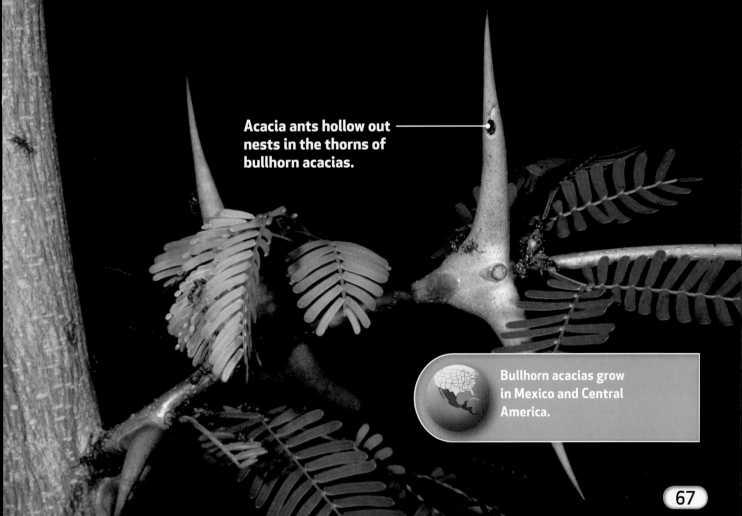

Acacia ants hollow out nests in the thorns of bullhorn acacias.

Bullhorn acacias grow in Mexico and Central America.

One-way Relationships

Some relationships help one organism, but neither help nor harm the other. Look at the fish attached to the shark's belly. The smaller fish are called remoras. They follow or hitch a ride with the shark so they can eat scraps of the shark's prey. The remoras don't help the shark, but they don't harm it either.

You can see this kind of symbiosis in almost any large tree. Chances are, the tree has at least one bird's nest. The nest benefits the birds but does not harm the tree.

Some plants live on trees without harming the trees. For example, tropical orchids live high up on trees where they can get plenty of sunlight. The orchids don't take water from the trees. Instead the orchids' rootlike structures take in water from the moist tropical air. Because the orchids grow close to the branches, the orchids do not compete with the trees for sunlight.

Remoras, also called suckerfish, use suction to attach themselves to sharks and rays.

Parasites The white object on this toad's head is a tick. It's swollen with blood sucked from the toad. The tick uses the blood as food. The tick is a parasite. A parasite is an organism that feeds off another organism, called the host. In this relationship, only the parasite benefits. The host is harmed.

A parasite usually does not kill its host. However, the parasite can weaken its host.

Many organisms live as parasites on others. Ticks live outside of animals' bodies. Other parasites, such as tapeworms, live inside their hosts. Disease-causing bacteria, protists, and fungi are parasites. Even some plants are parasites. Mistletoe is a plant that grows on trees. Rootlike structures on the mistletoe grow into the host tree branches, pulling water and nutrients from the tree.

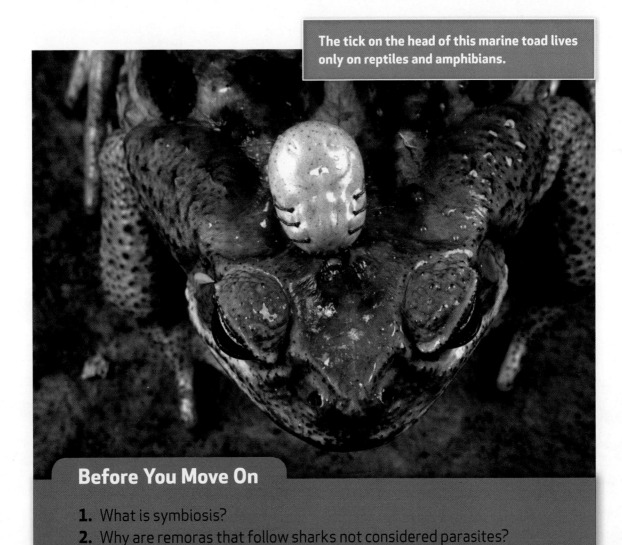

The tick on the head of this marine toad lives only on reptiles and amphibians.

Before You Move On

1. What is symbiosis?
2. Why are remoras that follow sharks not considered parasites?
3. **Infer** A parasite does not usually kill its host. How does having its host continue to live help a parasite survive?

Changing Communities

As organisms live in an ecosystem, they also change it. Think about how beavers affect the streams where they live. In the picture below, beavers have used mud and sticks to build a dam. The dam has turned a stream into a pond. The deeper water of the pond allows beavers to enter their dens underwater, which protects them from predators. This change helps the beavers survive.

The beaver dam affects other organisms in the community, too. Plants such as water lilies can grow in the still waters of the pond. Frogs, turtles, and sunfish also live there. But the trout that swam in the fast-flowing stream cannot live in the still water.

This beaver dam in Grand Teton National Park has created a pond that will one day become a meadow.

A beaver pond begins changing almost immediately. Mud and sand wash into the pond, making it shallower. Cattails and reeds take root in the shallow water. The pond has become a marsh. Red-winged blackbirds come to nest in the reeds. If the water becomes too shallow for the beavers, they move to a new area.

Eventually the marsh plants die and their remains fill the marsh. The land dries out. The rich soil from the decaying marsh plants now supports a meadow. The wildflowers and animals in the meadow are very different from those that lived in the pond or marsh.

The gradual replacement of one community by another over time is called succession. The changes in succession benefit some organisms. But other organisms die or move to new places.

A pond with many lily pads and other pond plants is in the middle stage of succession.

In the final stage of pond succession, a meadow of grasses and wildflowers forms.

Succession sometimes starts when a community is destroyed by natural events, such as a landslide or fire. In 1988, the weather in Yellowstone National Park was unusually dry. Then lightning started a fire that spread throughout much of the forest.

The fire killed thousands of trees and other plants. It killed many animals, too, and drove away others. The community in Yellowstone was changed—but more changes were to come.

Within a few weeks, life started returning to Yellowstone. The dead trees allowed large amounts of sunlight to reach the forest floor for the first time in many years. Grass seeds in the soil responded to the light and started growing. So did the seeds of fireweed and other wildflowers. Soon these plants covered large patches of the forest floor in carpets of green and pink.

 In the summer of 1988, fire destroyed much of Yellowstone National Park. The pictures below summarize how it's coming back to life.

Fireweed was one of the first plants that grew from the bare soil after the fire in Yellowstone.

The first trees to grow after the fire were lodgepole pines. Elk and other animals came to feed on the new plants.

As plants came back to Yellowstone, so did the animals. Ants and other insects had survived the fire in underground nests. Now they came out to feed on the growing plants. Birds returned to feed on the insects and plants. Elk and moose came back to browse on the grass and wildflowers.

Meanwhile, the fire had helped another species of plant grow. Many of the trees that burned were lodgepole pines. The intense heat caused the sealed cones of these pine trees to burst open. The seeds fell on the ground, and pine seedlings started to grow.

Today, young lodgepole pines are quickly replacing parts of the forest that burned. As the pines grow, they will block the sunlight from the forest floor. Sun-loving fireweed and grasses will die. Animals that live in pine forests, such as squirrels and owls, will replace animals that lived among the grasses and wildflowers.

Eventually succession may stop in an ecosystem. But that often takes hundreds of years. If the ecosystem is disturbed by fire, flood, or some other event, succession will begin again.

TECHTREK
myNGconnect.com

Enrichment Activities

As the lodgepole pines grow, they make up most of the trees in the forest community.

In two hundred years, many other kinds of trees will also grow in the forest, including spruce and fir trees.

Invasive Species In much of North America, marshes filled with purple loosestrife are a common sight. But 200 years ago, these plants did not exist in North America. European settlers brought them to add beauty to their gardens. The plants quickly spread across the continent. Now they're a problem.

Purple loosestrife crowds out cattails and other marsh plants that are food for fish, ducks, and other wildlife. But these animals don't eat purple loosestrife. This plant has no predators because it is not a part of the natural ecosystem. It's an invasive species.

Because invasive species usually have no predators in an ecosystem, their population grows quickly. The invasive species takes resources from the natural, or native, species. Often the native species die or move away.

A sea of purple loosestrife has taken over this marsh.

Cane toads were brought to Australia to eat insects that feed on crops. Today these toads are gobbling up native animals.

One invasive animal that has done much damage in the United States is the zebra mussel. This native of Russia was accidentally brought to North America by cargo ships. These ships take in water to stay balanced. The water from different parts of the world can contain eggs or larvae of different species. Water carrying the larvae of zebra mussels was released into the Great Lakes.

Zebra mussels have predators in Russia but not in the Great Lakes. They spread quickly throughout the Great Lakes and to other bodies of water. Zebra mussels eat huge numbers of tiny animals and plants. Because zebra mussels use the resources of other populations, such as fish, the natural populations die.

Zebra mussels are tiny mollusks that form clusters. These clusters can cover the sides of ships and clog water pipes.

Before You Move On

1. What is an invasive species?
2. What happens during the process of succession in an ecosystem?
3. **Predict** Fire ants were accidentally brought to the United States from South America by cargo ships. What will most likely happen to natural ecosystems after fire ants arrive?

Humans Change Ecosystems

The golden field below was once a wild prairie. The smaller picture shows how that prairie looked. It was filled with grasses and wildflowers. Those plants fed many kinds of animals. The interactions of all the different species formed a balanced ecosystem.

Today, farmers use this land to raise crops. Those crops feed many people. But the farm ecosystem is very different from the prairie.

Humans change the environment more than any other species. When people cut down trees for wood to build homes, they change forest ecosystems. When people build dams to make electricity, they change rivers. It would be impossible for people to meet their needs without changing ecosystems. But people are finding ways to preserve the natural world, too.

Wildflowers and grasses cover parts of the prairie that have not been plowed and planted for crops.

How have people changed the prairie ecosystem by farming?

How Humans Affect Land

Most of the food that people eat is grown on farms. It takes a great deal of land to grow enough food to feed the 6 billion people on Earth. Most of the land used for farms was once grasslands or forests.

Humans also need shelter. The land where people build cities and towns was once part of natural ecosystems. Roads, bridges, factories, and stores also use large amounts of land.

Many of the things that people use end up as trash. This trash is then dumped into landfills. The landfills take up space, too!

Today, people are working to restore natural areas. For example, landfills are being covered with soil and planted with grasses and wildflowers. Trees are being planted where forests have been cut down. New cities are being designed with places for gardens and trees. Large areas are being set aside as parks for wildlife.

When people throw away plastic and other trash, it usually ends up in a landfill.

How Humans Affect Water

From deserts to forests, water is a key part of every ecosystem. Water reaches land as rain and snow. It then flows from the land into rivers and streams. Finally it reaches the ocean.

Humans cannot live without drinking water. They also use water in farming, to generate electricity, and to carry away wastes. In many places, water from rivers is used to irrigate crops and lawns. Sometimes people take so much water that the organisms living in the rivers can no longer survive.

Some human activities pollute water. Anything that is added to the water that harms living things is a form of pollution.

Why does the water below look green? A blanket of algae covers the water. Fertilizer from nearby fields washed off the land and into the stream. Fertilizer makes algae grow faster, just as it makes crops grow faster. The blanket of algae blocks out the sunlight that underwater plants need. When the algae die, they decay. The process of decay uses up much of the oxygen in the water that fish need to live.

Fertilizer washing off fields caused this quick growth of algae, called an algal bloom.

Pesticides, chemicals from factories, and wastes from humans and livestock also pollute water. Pesticides are used to kill insects that eat crops. When pesticides wash into streams and lakes, they harm the animals living there.

Do you think the ocean is too big to be polluted? In fact, most forms of pollution eventually flow into the ocean. This pollution harms many ocean ecosystems. Today, a huge patch of plastic trash floats in the middle of the Pacific Ocean. Animals are killed when they mistake the plastic for food or become entangled in the trash.

Recently the United States has made progress in cleaning up water pollution. Lake Erie used to be one of the most polluted lakes in the world. But people worked hard to clean it up. Their efforts are a great success story of saving ecosystems.

Until the 1970s, waste from factories, such as the pulp from this paper mill, flowed into Lake Erie. Truckloads of trash were also dumped into the lake.

Today, the water in Lake Erie is clean enough for fish to live and for humans to enjoy.

How Humans Affect Air All animals that live on land need clean air to breathe. But many human activities pollute the air. Air pollution harms both humans and wildlife.

Most air pollution comes from burning fuels, such as wood, coal, and gasoline. You can see the ash and tiny bits of material that come from some fires. Burning also gives off many gases that you cannot see.

The haze below is a common sight in many cities. It's smog. This kind of air pollution forms when sunlight reacts with gases given off by cars and trucks.

You may have heard a weather reporter give a "smog alert." During those times, the amount of smog in the air is dangerous for people who have asthma or other breathing problems.

Cars and trucks are a major source of air pollution in American cities.

The purple haze that covers Los Angeles on this day is smog.

Another kind of pollution comes from power plants that burn coal and oil to make electricity. As they burn, these fuels give off tiny drops of acid. This acid falls to Earth when it rains. If acid rain falls into lakes, it makes them more acidic. Then many fish and their eggs cannot survive. Acid rain also affects soil and trees. Sometimes it can kill entire forests.

Whenever people burn fuels, carbon dioxide goes into the air. Most scientists think that this carbon dioxide is changing Earth's climate. A changing climate will affect nearly all ecosystems on Earth.

Today, nations around the world are working to clean up the air. New technologies let cars and factories release less pollution. One of the best solutions is to burn less fuel.

Acid rain from factories killed this forest in the Czech Republic.

Helping the Environment

People around the world are working to improve the environment. You can too. Instead of traveling in cars, you can walk, bike, or ride the bus. This will reduce air pollution and save energy.

You can reduce the amount of resources you use. For example, you can use less plastic by drinking tap water instead of bottled water. Many people take cloth bags to the grocery store. This cuts down on the number of plastic bags that are made and thrown away.

Planting trees is a good way to help the environment.

Riding a bicycle to school saves energy and reduces pollution.

Recycling saves resources. Maybe you already recycle paper and plastic at home. It's amazing how much less trash you have when you recycle. Do you have a compost pile where you put leaves and kitchen wastes? Compost can help a garden grow.

If you look, you will find many ways to help the environment. You may be able to participate in tree-planting activities. Maybe you can help pick up trash. Your actions will help improve the health of your local ecosystem.

Science in a Snap! Personal Impact

ACTIVITY	TALLY
Recycle	
Turn off lights	
Turn off water	
Ride a bike or walk	
Pick up trash	

How are you doing in protecting the environment? Complete a checklist like the one above to find out. How many times did you do each of these things this week?

What other things could you do to help keep ecosystems healthy?

Before You Move On

1. What are three kinds of water pollution? Where does each come from?
2. Describe how burning fuels such as gasoline and coal can affect the air.
3. **Generalize** Like all living things, humans need food, water, and shelter. How does the human use of these resources affect other organisms? Use the word *competition* in your answer.

NATIONAL GEOGRAPHIC

ROOFTOP GARDENS

Do you know what the roof of your school looks like? Chances are, it's not much to look at. Why should it be? It's just a roof. Well, people are realizing that roofs can do a lot more than just keep the rain out. They can help the environment.

Most cities are built on land that was once part of a forest, prairie, or desert ecosystem. The plants that covered the ground are long gone. That's too bad, because plants are very helpful.

In summer, the garden on the roof of Chicago's City Hall is a pleasant place to take a break. Before the garden was planted, temperatures sometimes reached 65°C (150°F).

Besides providing food and homes for animals, plants give off oxygen and cool the air. Most people agree that they're attractive too. Large areas of plants can make a city more livable. But where can you grow plants in a place covered with buildings?

One answer comes from an old idea. Settlers on the prairies used sod—slabs of grass and soil—to cover their roofs. These "living roofs" kept the homes cool in the summer and warm in the winter. Today, many cities are using the same idea.

Living roofs have many advantages. They reduce the costs of heating and cooling buildings. They capture rainwater that would otherwise wash into the sewers. Other living roofs are planted to attract birds. These roofs don't replace the original ecosystems, but they help improve city life for people as well as animals.

Green roofs like this one in New York City attract birds and other wildlife.

Conclusion

Ecosystems are made up of communities of living things and the nonliving things in their environment. The organisms in an ecosystem interact in many different ways. These interactions include predator-prey relationships and competition for resources. Some species rely on other species through symbiosis. Each kind of living thing has a niche, its role in the ecosystem. The actions of organisms often change their environment. As conditions in the environment change, one community may be replaced by another. This process is called succession. Humans change ecosystems more than any other species on Earth.

Big Idea Ecosystems are made up of many different kinds of interactions among living things and their environment.

Vocabulary Review

Match the following terms with the correct definition.

A. symbiosis
B. population
C. ecosystem
D. niche
E. succession
F. community

1. The gradual change in living communities over time
2. An organism's role in its ecosystem
3. All the members of a species in an area
4. All the different species living and interacting in an ecosystem
5. The close association of different species that live together
6. All the living and nonliving things in an area and their interactions

Big Idea Review

1. **List** What are the three different kinds of symbiosis? Give an example of each kind.

2. **Recall** Why are zebra mussels a problem in the Great Lakes area?

3. **Explain** How are plants able to begin growing in an area after a forest fire?

4. **Contrast** How is a population different from a community?

5. **Predict** Many people are moving into deserts. How might this movement affect the ecosystems of the rivers that flow through these deserts?

6. **Make Judgments** Your class has been asked to help improve the natural environment of your community. What are three things that you and your class could do?

Write About Interactions

Analyze This beaver is cutting down trees to dam a stream. How will the actions of the beaver affect the other organisms in its community?

NATIONAL GEOGRAPHIC

CHAPTER 2 LIFE SCIENCE EXPERT: CONSERVATIONIST

If you want to help save species, you may want to follow in the footsteps of Milagros López-Mendilaharsu. Dr. López-Mendilaharsu is a researcher who works at the headquarters of the Brazilian Sea Turtle Conservation Program.

What do you do as a conservationist?

I study the biology and ecology of sea turtles. They have been on Earth for over 100 million years. Now, they are endangered or at risk of disappearing from Earth. I use knowledge from my research to protect them and to help keep marine ecosystems healthy. I also help everyone learn how to preserve our oceans and coastal habitats.

Milagros López-Mendilaharsu helps protect sea turtles and ocean ecosystems.

What's a typical day like for you?

I get up very early and join the technical team. We go to a place where the turtles are eating and perform in-water research activities. Then I return to the office to update records, analyze data, and do scientific writing. In the afternoon, I go to meetings where we discuss ongoing research on sea turtles, nesting beaches, and coastal habitats. During nesting season, I do night patrols on the beach to observe females laying eggs.

What did you like best about school?

I remember I liked biology, especially the study of animals and the environments in which they live. Also I enjoyed conducting experiments and doing field work to understand the natural world.

TECHTREK myNGconnect.com
Digital Library

Milagros watches a loggerhead sea turtle lay eggs at night.

What is the coolest part of your job?

Traveling to exotic places and helping conserve marine resources are really cool. My favorite experience was diving with a leatherback, the largest sea turtle, and seeing her swim back to the deep blue ocean.

What was the biggest factor in choosing your career?

I chose my career to help protect wildlife and their environment. Knowing that your contribution can aid in the survival of marine species is priceless.

What advice would you give young people who want to do what you do?

First of all, they need to be fascinated by nature. To be successful, they need to have passion, to be determined, and to be patient.

Milagros and a coworker must scuba dive to take measurements of a hawksbill sea turtle.

NATIONAL GEOGRAPHIC

BECOME AN EXPERT

Interactions in a Coral Reef

Coral reefs are like great cities in the ocean. They are home to an amazing community of living things. Clams, sponges, and seaweeds grow from the reef. Sea stars and shrimp creep over the surface, hunting for food. Sea turtles, sharks, and colorful fish swim among the towering corals. No other marine ecosystem has so many different species.

Coral reefs can grow only in certain places. First, the ocean water must be warm. That's why most coral reefs are in the tropics. Second, the water must be shallow and clear so plenty of sunlight can reach the bottom. Third, there must be waves to bring oxygen and other nutrients to the reef. Less than one percent of the ocean has conditions that are just right for coral reefs to grow.

community
A **community** is all the different organisms that live and interact in an area.

ecosystem
An **ecosystem** is all the living and nonliving things in an area and their interactions.

Coral animals grow in colonies. Each coral animal is surrounded by a hard, stony skeleton. The skeletons of the individual corals grow together. If you look closely, you can see that one piece of coral is really a population of hundreds of coral animals.

When coral animals die, their hard skeletons remain. Soon other corals grow on top of them. Over time, the skeletons of millions of tiny coral animals form a reef.

The body of a coral animal has a mouth and stinging tentacles. Each coral animal is surrounded by a hard skeleton.

tentacle
mouth
coral skeleton

The large stony structures of a reef are made by tiny coral animals.

population
A **population** is all the members of one species living in a certain area.

BECOME AN EXPERT

Symbiosis on the Reef The body of a coral animal is colorless. So why are coral reefs so colorful? Their color comes from tiny algae that live inside their bodies. The algae and coral are connected by **symbiosis**. Neither organism can live without the other.

Corals get some energy by eating other animals. But most of their food is made by the algae in their bodies. The algae make this food by photosynthesis.

In return, the bodies of the coral animals give the algae a safe place to live. The coral animals also give off carbon dioxide, which the algae need for photosynthesis.

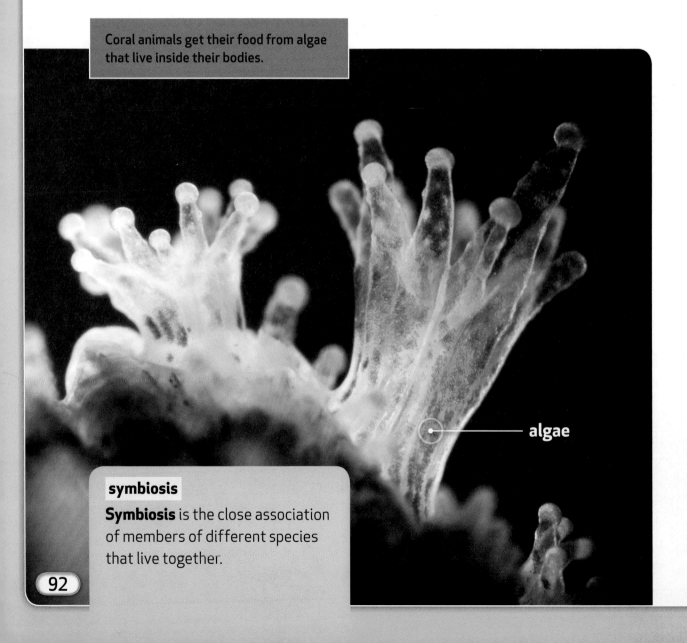

Coral animals get their food from algae that live inside their bodies.

algae

symbiosis
Symbiosis is the close association of members of different species that live together.

Many organisms on the reef have symbiotic relationships. One of the most remarkable is the partnership of the clownfish and sea anemone.

Clownfish are usually seen swimming among the tentacles of sea anemones. These stinging tentacles are deadly to many fish.

But clownfish have a slimy covering that protects them from the tentacles' poison. Staying close to sea anemones keeps the clownfish safe from larger fish.

Sea anemones are helped by this relationship, too. The clownfish keep the anemones clean by eating parasites and dead skin. The clownfish also chase away animals that might eat the sea anemones.

The stinging tentacles of the sea anemone protect the clownfish from predators.

BECOME AN EXPERT

How Reefs Form When a new volcanic island rises out of a tropical sea, a coral reef may start to form. The shallow water around volcanic islands often has the sunlight, air, and nutrients that corals need to grow.

Coral reefs develop through a process of succession. As the coral animals start to grow, they change their environment in many ways.

The physical changes in the reef make it possible for new species to live there. As the reef gets larger, the number and kinds of organisms living there increase. When seaweeds attach to the coral, snails that graze on the seaweeds can also live on the reef. Soon animals that eat the snails move in, too. Thus, new communities take the place of others as conditions change.

GROWTH OF A CORAL REEF

NEW ROCK A new reef begins when coral animals attach themselves to a hard surface, such as a new lava flow.

CORAL COLONIES GROW Colonies of coral grow upward from the rocks. Sponges, seaweeds, fishes, and other animals join the new community.

succession
Succession is the gradual replacement of one community by another over time.

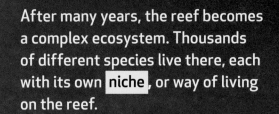

After many years, the reef becomes a complex ecosystem. Thousands of different species live there, each with its own niche, or way of living on the reef.

A mature coral reef takes hundreds of years to form. A colony of coral may grow only a few millimeters per year—less than a pencil's thickness. Others grow more quickly, as much as 20 centimeters (about 8 inches) per year. Many of the living reefs on Earth today are between 5,000 and 10,000 years old.

MATURE REEF As colonies of coral build upon one another, the reef grows taller and wider. The mature reef is home to an incredible number of species.

niche

A **niche** is the way an organism interacts with living and nonliving parts of its ecosystem.

BECOME AN EXPERT

Reefs in Danger Coral reefs have been around for millions of years. But they are also very delicate. Scientists everywhere agree that coral reefs are in danger. More than a quarter of Earth's coral reefs have disappeared. Human activities are the biggest threat to their survival.

Coral Bleaching As Earth's atmosphere gets warmer, so does the ocean. Corals are very sensitive to changes in water temperature. If the water gets too warm, the coral animals push the algae out of their bodies. As a result, the corals turn white. This process is called coral bleaching. Without their symbiotic partners to feed them, the corals die.

Fishing Practices Many people who live in tropical regions rely on fish that live on coral reefs for food. But some modern ways of catching fish damage the reefs. One destructive practice is bottom trawling. A trawl is a large net pulled by a boat to catch fish. In bottom trawling, the net is rolled across the ocean floor on huge wheels. When one of these nets rolls across a reef, the delicate corals are crushed.

Coral bleaching takes place when the ocean water gets too warm.

The anchor chain of a large ship has broken apart this coral reef.

Pollution What people do on land affects coral reefs. When people apply too much fertilizer to crops and lawns, it runs off the land into rivers. Eventually it reaches the ocean. Sewage also flows into the ocean.

The nutrients in fertilizers and sewage cause some kinds of algae to increase greatly. These algae block sunlight and use up most of the oxygen in the water. This can kill the coral.

Building on Land When people build cities and homes near the shore, it affects nearby reefs. Mud from construction sites washes into the ocean. As the water becomes muddier, less sunlight reaches the reefs. The algae in the coral cannot make food, so the coral animals starve.

Tropical Fish Trade Many people keep tropical fish in aquariums. These fish are often collected by spraying poison on coral reefs. The poison stuns the fish so they can be collected. But it also poisons other animals on the reef.

Tourism Every year many divers visit coral reefs to see the beautiful creatures that live there. Sometimes these divers harm the reefs accidentally. Bumping into the reefs can weaken or break the coral. A few divers break off pieces of coral to take home as souvenirs. Other people buy coral in gift shops. Since coral grows very slowly, it should never be taken from a reef.

Careless divers can damage reefs. Bubbles from scuba tanks can stir up mud and block out light.

BECOME AN EXPERT

Protecting and Restoring Coral Reefs Many people are working to save Earth's coral reefs. Educating people about the importance of coral reefs is one important step. Passing stronger laws to protect them is another.

Much is also being done to restore coral reefs. Where reefs are damaged, artificial reefs can be built. Artificial reefs have solid surfaces on which corals can grow. One way of making an artificial reef is to sink large objects such as old cars and ships in areas where reefs can form. Reef balls made of concrete and fiberglass are another type of artificial reef.

Corals are beginning to grow on these reef balls in the Caribbean Sea. Holes in the reef balls let the ocean's natural currents move through them.

Would you like to help save the reefs? Even if you don't live near the ocean, there several things you can do.

Don't pollute. Never leave trash on the beach or put garbage in the water. Avoid putting too much fertilizer or pesticide on your garden or lawn. Eventually these chemicals wash into the ocean.

If you buy fish for a saltwater aquarium, be sure that they were collected responsibly. Never buy living coral to put in your aquarium.

If you ever get the chance to snorkel or dive near a reef, be very careful. Never touch the reef. Don't stir up mud from the ocean floor that could smother the corals. Just be glad you are seeing such an amazing ecosystem!

This diver is attaching corals to an artificial reef.

Coral and sponges grow on an artificial reef near Bali, Indonesia.

BECOME AN EXPERT

CHAPTER 2
SHARE AND COMPARE

Turn and Talk What are the living and nonliving parts of the reef ecosystem? Form a complete answer to this question with a partner.

Read Select two pages in this section. Practice reading the pages. Then read them aloud to a partner. Talk about why the pages are interesting.

Write Write a conclusion about the interactions in coral reefs. In your conclusion, state what you think is the Big Idea of this section. Share what you wrote with a classmate. Compare your conclusions. Did you include interactions with both living and nonliving parts of the ecosystem?

Draw Draw a picture of an organism that lives on a coral reef. Identify the ways that this organism interacts with other living things in the reef community. Post your drawing in the classroom.

LIFE SCIENCE

After reading Chapter 3, you will be able to:

- Recognize that all living things need energy to survive. **PRODUCERS AND CONSUMERS**
- Recognize that different ecosystems contain different organisms, but energy always flows through the ecosystem from one organism to another.
  PRODUCERS AND CONSUMERS, HERBIVORES, CARNIVORES, AND OMNIVORES, DECOMPOSERS, FOOD CHAINS AND FOOD WEBS
- Recognize almost all food eaten by animals can be traced back to green plants.
  PRODUCERS AND CONSUMERS
- Describe the process of photosynthesis. **PRODUCERS AND CONSUMERS**
- Recognize that plants (producers) can make their own food, but animals (consumers) cannot make their own food and must eat plants and other animals.
  PRODUCERS AND CONSUMERS, HERBIVORES, CARNIVORES, AND OMNIVORES
- Recognize that food chains and food webs are processes by which energy moves through an ecosystem. **FOOD CHAINS AND FOOD WEBS**
- **Science in a Snap!** Recognize that plants (producers) can make their own food, but animals (consumers) cannot make their own food and must eat plants and other animals.
  HERBIVORES, CARNIVORES, AND OMNIVORES

CHAPTER
3

HOW DOES ENERGY MOVE IN AN

An orangutan munches plants found in its rain forest home. Orangutans need energy to survive. Energy moves through this rain forest ecosystem as plants make food and animals eat plants or other animals.

MOVE ECOSYSTEM?

This orangutan is eating fruit to get the nutrients it needs.

SCIENCE VOCABULARY

chlorophyll (KLOR-uh-fil)

Chlorophyll is the green material in plants that absorbs sunlight, which the plant uses to make food. (p. 108)

These trees use chlorophyll to absorb sunlight.

photosynthesis (FŌ-tō-SIN-thuh-sis)

Photosynthesis is the process by which green plants make food by using energy from sunlight. (p. 108)

Plants make their food through the process of photosynthesis.

my Science Vocabulary

chlorophyll (KLOR-uh-fil)
food web (FŪD WEB)
food chain (FŪD CHĀN)
photosynthesis (FŌ-tō-SIN-thuh-sis)

food chain (FŪD CHĀN)

A **food chain** is a process by which energy passes from one living thing to another. (p. 116)

PRODUCER sagebrush

HERBIVORE lubber grasshopper

The first organism in a food chain is always a producer.

food web (FŪD WEB)

A **food web** is a process that combines many food chains to show how energy moves through an ecosystem. (p. 118)

This food web shows that one predator might eat many kinds of prey.

105

Producers and Consumers

A sunflower needs energy to make seeds. A grasshopper needs energy to hop from one blade of grass to another. A cheetah needs energy to chase down a gazelle. All living things need energy, but different organisms get the energy they need in different ways.

Plants and algae are called producers. They use energy from sunlight to make their own food. This food gives them the energy they need to live and grow.

The algae on the pond, and the green plants that grow beside it, produce food that many other organisms use for energy.

Animals and some other organisms cannot produce their own food. They are called consumers. They get food by consuming, or eating, other organisms. Some consumers, such as cows, get energy directly from producers by eating grass and other plants. Other consumers, such as birds, get energy by eating other animals.

All animals need the energy in food to live and grow, and almost all the energy in the food that animals eat can be traced back to the energy in plants.

These cows get their energy from eating grass.

Imagine taking a walk through a forest on a spring day. You may notice the many different colors of the flowers around you. You may also notice that most of the plants are green. The reason that the plants are green is that they contain chlorophyll. Chlorophyll is the green material in plants' leaves that absorbs sunlight. Plants absorb light energy from the sun, and combine it with carbon dioxide and water to make food. This process is called photosynthesis. In photosynthesis, the plant makes sugars. The plant then uses the sugars for food.

PHOTOSYNTHESIS

sunlight

sugars

water

Chlorophyll gives these plants their green color.

Photosynthesis brings energy into an ecosystem. Animals can't use light energy directly from the sun as plants do. Plants change the light from the sun into energy that is consumed and used by animals and other living things.

Photosynthesis is important in another way too. To make food, plants take carbon dioxide out of the air and put oxygen back into the air. Animals need oxygen to live. When they breathe, they take oxygen from the air and put carbon dioxide back into the air. Then plants use that carbon dioxide to make more food.

Oxygen bubbles form on these *Elodea* plants as they recycle oxygen back into the ecosystem.

Before You Move On

1. What is photosynthesis?
2. Why are producers important to consumers?
3. **Draw Conclusions** What would happen to the animals if all the plants in their ecosystem died?

Herbivores, Carnivores, and Omnivores

All the animals on Earth are consumers. Consumers can be grouped by what they eat. What kinds of food do consumers eat? Some eat plants. Some eat other animals. Some eat both plants and animals.

An herbivore is a consumer that gets its energy from eating plants. Look at the insect in the photo below. It is getting energy from the plant it is eating. The insect will use the energy it gets from the plant to live and grow.

This common cockchafer is an herbivore because it eats only plants.

Carnivores Consumers that eat other animals are called carnivores. The shark shown below is a carnivore. It gets its energy from eating other animals.

Some carnivores, such as vultures, don't kill animals to get food. They eat what remains of animals that have already died or have been killed by other carnivores. Vultures are a kind of carnivore called scavengers.

A shark gets its energy by eating another animal. The shark is a carnivore.

Omnivores Consumers that eat both plants and animals are called omnivores. Many animals, including pigs, seagulls, cockroaches, and chimpanzees, are omnivores.

Look at the photo below. The black bear in the photo is an omnivore. Its diet is made up of plant parts, such as young grasses, fruits, nuts, and berries. Black bears also eat insects, fish, and rabbits. Black bears may even kill and eat young deer and elk. Black bears eat both plants and animals to get the energy they need to live and grow.

This black bear has caught a fish. It also eats berries and other kinds of fruit.

This raccoon is an omnivore.

Raccoons are omnivores too. Raccoons eat many different kinds of fruit and other plant parts. They also eat small animals or eggs.

Raccoons live in many different places, including places where people live. People throw away large amounts of food, and raccoons that live nearby eat the food. These raccoons' ability to eat many different kinds of food has helped them survive and thrive in their environment.

Science in a Snap! What Have You Eaten?

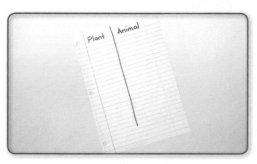

On a sheet of paper, create a two-column chart. Label one column *Plant*, and the other column *Animal*.

Think of everything you have eaten for the last two days. Decide if it comes from a plant or an animal, and record it in the chart.

Did you record more foods from plants or from animals? Do you notice any patterns?

Before You Move On

1. What are the three kinds of consumers?
2. How are herbivores like omnivores? How are they different?
3. **Apply** Wolves and black bears both eat rabbits. Wolves are carnivores and bears are omnivores. Would wolves or bears be more likely to survive in an ecosystem if the rabbit population died out? Explain why.

Decomposers

Energy moves from sunlight to plants to herbivores, carnivores, and omnivores. Energy also moves through decomposers. Decomposers are living things that break down animal skins, tree trunks, and other decaying materials. They get the last bit of energy from plants and animals.

One example of a decomposer is mold. You may have seen a fuzzy mold growing on a piece of stale bread. The mold breaks down the bread and gets the energy it needs to live and reproduce.

These mushrooms help to break down the log they are growing on.

If you hike in the woods, you might see mushrooms on a decaying log. Actually, the mushrooms and another living thing, bacteria that you cannot see, are working together to break down the log. Bacteria are one-celled organisms that live in every habitat on Earth—in soil, in water, in air, and even in other organisms. Bacteria break down dead or decaying organisms, such as a log. They release the nutrients from those organisms into the soil. Then these nutrients are used by plants.

These bacteria help return nutrients to the soil. They are magnified many thousands of times.

Before You Move On

1. What is a decomposer?
2. How is a decomposer like an herbivore or carnivore? How is it different?
3. **Generalize** How do decomposers help plants? How are they helped by plants?

Food Chains and Food Webs

Each organism plays an important part in its environment. You already know that every living thing needs energy to survive. Some living things get energy directly from the sun. Some get energy from eating other organisms.

A **food chain** is a process by which energy passes from one living thing to another. In the food chain shown here, you can trace one path of energy through a desert ecosystem.

DESERT FOOD CHAIN

This diagram shows a food chain of desert organisms. Food chains from different ecosystems follow the same pattern as this food chain of desert organisms.

ENERGY SOURCE
sun

PRODUCER
sagebrush

HERBIVORE
lubber grasshopper

The first organism in this desert food chain is sagebrush, a producer. The first organism in a food chain is always a producer. Producers are the living things that bring energy into the ecosystem.

After the sagebrush, energy then flows to the consumers. The lubber grasshopper eats the sagebrush, and the Texas horned lizard eats the grasshopper. Finally, the red-tailed hawk eats the Texas horned lizard. The sun's energy has moved through this entire desert food chain.

CARNIVORE
Texas horned lizard

CARNIVORE
red-tailed hawk

Look at the food web below. Imagine what would happen if a disease caused many of the mice to die. Plants would thrive without mice to consume them. Badgers could die without the mice to eat. If one thing is changed in an ecosystem, the ecosystem may change drastically.

What if a new insect were released in a desert? If the new insect is one that desert predators don't eat, that insect might multiply and devour most of the plants. Herbivores would die without plants to eat. Carnivores and omnivores would die without the herbivores to eat.

Before You Move On

1. What is a food chain?
2. Why doesn't a food chain give a complete picture of the flow of energy in an ecosystem?
3. **Draw Conclusions** Suppose a disease killed most of the red-tailed hawks in a desert area. What effect would that have on the area's animal life?

NATIONAL GEOGRAPHIC
DEEP-SEA VENTS
LIVING ON THE EDGE

All the living things around you depend on sunlight for their energy. Can life exist on the dark ocean floor where no sunlight ever reaches?

Much of the ocean floor cannot support life. However, scientists have found some special areas called deep-sea vents. At these vents, water heated by hot, melted rock beneath the ocean floor escapes through cracks in the rock. This heated water contains chemicals. Certain kinds of bacteria use these chemicals to make their own food. These bacteria are the producers for food webs in the deep ocean. The bacteria support many consumers, including giant tube worms, fish, crabs, and giant clams.

Sea stars looking for food near a deep-sea vent.

Giant tube worms grow on a deep-sea vent 2,620 m (8,600 ft) below the surface.

Warm deep-sea vents can be separated by wide areas of cold ocean floor. These areas cannot support living things. How do organisms move through these areas and find sea vent ecosystems that contain food? Marine ecologists are exploring these and other questions. To find answers, they dive several kilometers or deeper in small underwater crafts, such as *Alvin*. Scientists have been visiting the Galápagos Rift vents in the Pacific Ocean since 1977. To this day, they continue to find new kinds of living things that have never been seen before.

Scientists use *Alvin*, a small submarine called a submersible, to explore ecosystems deep in the ocean.

Two fish swim past a deep-sea vent looking for food.

Conclusion

All living things need energy to survive. The energy used by most living things comes from the sun. Through photosynthesis, plants use energy in sunlight to make food. These producers are eaten by consumers, which get energy from the food. Consumers can be classified as herbivores, carnivores, or omnivores. Decomposers break down dead organisms and recycle the nutrients back into the soil. Energy spreads through an ecosystem as one organism consumes another organism. This flow of energy can be described by a food chain or a food web.

Big Idea Energy moves through an ecosystem as producers make food and are then consumed by other organisms.

Vocabulary Review

Match each of the following terms with the correct definition.

A. chlorophyll
B. photosynthesis
C. food chain
D. food web

1. A process by which energy passes from one living thing to another
2. The green material in plants that absorbs sunlight, which the plant uses to make food
3. The process by which green plants make food by using energy from sunlight
4. A process that combines many food chains to show how energy spreads through an ecosystem

Big Idea Review

1. **Define** What is a consumer?

2. **Identify** What do green plants need to make food?

3. **Explain** What is a decomposer's role in an ecosystem?

4. **Classify** Opossums eat many kinds of food, including insects, small mammals, and fruit. What kind of consumer is an opossum?

5. **Analyze** Deep-sea vents eventually fill with hardened rock to become solid ocean floor. How would that cause the ecosystem to change?

6. **Draw Conclusions** Lions eat zebras. How might the lion population be affected if the zebra population died out?

Write About Energy in an Ecosystem

Explain Use the diagram to tell about one path of energy in an ecosystem. Use words that tell about producers, consumers, predators, and prey.

NATIONAL GEOGRAPHIC

CHAPTER 3 LIFE SCIENCE EXPERT: ECOLOGIST

Neo Martinez, Ecologist

"It's kind of like detective work," says ecologist Neo Martinez. "To discover something new about nature, you have to look really closely for a long time without being fooled." Martinez is a Director of the Pacific Ecoinformatics and Computational Ecology Lab. He works with other scientists to explore how different species interact in their ecosystems. Then the scientists all work together to try and figure out why the different species interact in various ways.

Martinez says, "It's amazing to discover something new about nature that no one has thought of or seen before." He explains, "My lab has come up with a big theory about what food webs look like and how they function. It allows us to predict what happens when species go extinct and when new species invade an ecosystem. Most ecologists thought that ecosystems were too complex to be able to do that. But I think we're proving those ecologists wrong."

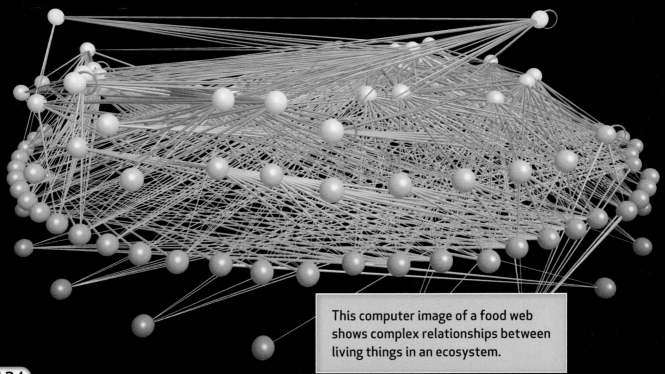

This computer image of a food web shows complex relationships between living things in an ecosystem.

Food Webs You eat a variety of foods, and so do other living things. Most insects eat different kinds of plants. A snake may eat mice, grasshoppers, and other kinds of animals. A  is a process that combines many food chains to show how energy moves through an ecosystem. A food web also shows how different organisms depend on one another.

In the food web below, you can see how living things in a ecosystem depend on one another for food. Like food chains, food webs show how the energy in an ecosystem begins with producers. From there, energy spreads out through a network of herbivores, carnivores, and omnivores. Unlike food chains, however, food webs show that living things eat a variety of organisms.

DESERT FOOD WEB

As director, Martinez helps decide what issues his lab will explore. He also makes sure his lab has everything it needs to continue its work. He spends most of his time writing, traveling, giving talks, and staying connected with scientists around the world.

Martinez did not always dream of being an ecologist. When he was younger, he wanted to be a veterinarian. But after working, biking, and skiing around the country, he decided that it was more important to find ways to take care of the planet.

Martinez says, "I hope my biggest impact is helping people to understand how much all organisms, including humans, depend on each other for the food we eat, the air we breathe, and the life we live. I think understanding that will help us to be nicer to each other and the planet. That will make life better for everyone."

Neo Martinez works in his lab to learn more about how energy moves in ecosystems.

NATIONAL GEOGRAPHIC

BECOME AN EXPERT

Energy in Ecosystems: The Arctic

The Arctic The organisms in an ecosystem are connected to one another through different food webs . Although food webs in an ecosystem connect different kinds of living things, all food webs show how energy moves through the ecosystem as one organism consumes another.

This polar bear hunts marine mammals, such as seals, but may also scavenge dead fish or other animals.

food web

A **food web** is a process that combines many food chains to show how energy spreads through an ecosystem.

Picture a lone polar bear hunting for food on a cold, white stretch of ice on the Arctic Ocean. Polar bears are at the end of the **food chain**. Which organisms are at the beginning? Recall that the first organism in a food chain is always a producer. No plants grow on the ice. The producers in this cold environment live in the ocean. Phytoplankton, such as those shown to the right, are tiny organisms that make their own food through **photosynthesis**. It's amazing to think that these microscopic organisms form the beginning of a food web that supports large animals such as whales, walruses, and polar bears.

Polar bears do not eat phytoplankton directly. They are carnivores and eat seals, young walruses, and other animals. Seals and walruses eat cod, salmon, and other fish. Yet all of these organisms can trace their energy back to the phytoplankton.

These phytoplankton make their own food.

food chain
A **food chain** is a process by which energy passes from one living thing to another.

photosynthesis
Photosynthesis is the process by which green plants make food by using energy from sunlight.

BECOME AN EXPERT

In the diagram below, trace how energy moves through an Arctic food chain. Phytoplankton bring energy into the ecosystem through photosynthesis. The phytoplankton do not use **chlorophyll**, like green plants do, to make their food, but they still use energy from the sun. The phytoplankton is consumed by tiny, floating animals called zooplankton, which include krill and many other small organisms. Zooplankton pass on their energy as they are consumed by fish, clams, and other animals.

ARCTIC FOOD CHAIN

phytoplankton

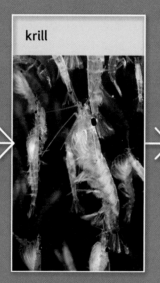
krill

Atlantic cod

harp seal

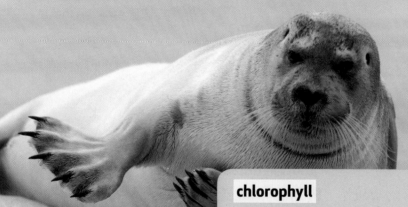

chlorophyll
Chlorophyll is the green material in plants that absorbs sunlight, which the plant uses to make food.

You might think that a large baleen whale would need to eat large animals to get enough energy to survive. Actually, baleen whales live mostly on zooplankton. Baleen whales don't have teeth. Instead, they have big plates that work like a filter. These whales take in huge amounts of water and squeeze it out through the plates, trapping krill and other tiny animal life.

Beluga whales, also found in Arctic waters, do have teeth. That's a clue that they eat a different kind of food than baleen whales eat. Belugas hunt fish, squid, crabs, and other prey. But sometimes they find themselves prey to a polar bear!

The filtering plates of a baleen whale are made out of a material like horses' hooves—or your fingernails!

These zooplankton look like tiny shrimp.

BECOME AN EXPERT

The diagram below shows an Arctic food web. This food web combines many food chains that energy passes through in the Arctic ecosystem. Energy from the sun is converted to food by the phytoplankton. This energy is passed through many different consumers such as fish, seals, polar bears, and whales.

Unlike in a food chain, however, many different connections between animals that live in the Arctic ecosystem can be seen in a food web. The food web shows the complex flow of energy between various living things. The Arctic food web also shows how delicate the interactions of the animals in the ecosystem can be.

ARCTIC FOOD WEB

Small changes in a food web can cause dramatic changes to the energy flow in an ecosystem. If one part of the food web is removed, it could make all other living things in the ecosystem suffer. For example, if a certain kind of fish died out because the water was too warm, many seals would also begin to die.

This would, in turn, affect the polar bears, who depend on the seals for food.

Scientists are trying to keep climate changes to the Arctic ecosystem to a minimum, but human activity in the area, as well as pollution, have made changes to the Arctic ecosystem that may be irreversible.

BECOME AN EXPERT

CHAPTER 3
SHARE AND COMPARE

Turn and Talk How does energy move through the Arctic ecosystem? Form a complete answer to this question together with a partner.

Read Select two pages in this section. Practice reading the pages. Then read them aloud to a partner. Talk about why the pages are interesting.

Write Write a conclusion that summarizes what you have learned about how energy flows through the Arctic ecosystem. In your conclusion, restate what you think is the Big Idea of this section. Share what you wrote with a classmate. Compare what each of you wrote. Did you recall that many different kinds of organisms interact with each other in the Arctic?

Draw Form groups of four. Have each person draw a different organism from the same food chain. Label the pictures. Put the drawings in order to show how energy flows through your food chain.

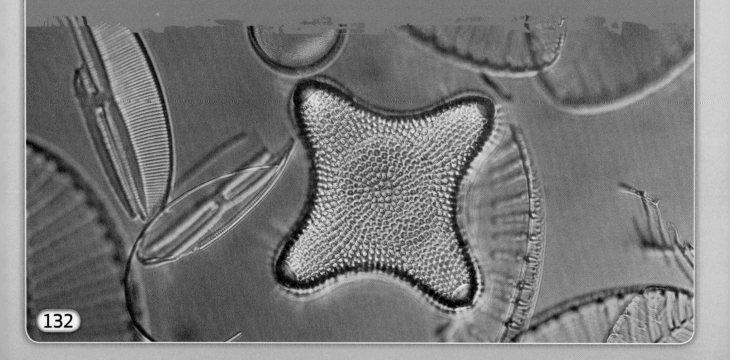

LIFE SCIENCE

After reading Chapter 4, you will be able to:

- Identify physical characteristics that help plants and animals survive in their environment. **PHYSICAL CHARACTERISTICS OF LIVING THINGS**

- Distinguish between inherited and acquired characteristics of living things.
  PHYSICAL CHARACTERISTICS OF LIVING THINGS

- Explain how behaviors help animals survive in different environments.
  BEHAVIORS HELP ANIMALS SURVIVE

- Distinguish between instinct and learned behavior. **BEHAVIORS HELP ANIMALS SURVIVE**

- Explain how communication helps animals survive and reproduce.
  BEHAVIORS HELP ANIMALS SURVIVE

- Identify adaptations and behaviors that help organisms survive in changing seasons.
  LIFE CYCLE ADAPTATIONS

- Explain how differences among individuals allow some species to adapt to environmental change, while others go extinct. **WHEN ENVIRONMENTS CHANGE**

- Describe how fossils provide evidence about organisms that lived long ago.
  WHEN ENVIRONMENTS CHANGE

- *Science in a Snap!* Identify physical characteristics that help animals survive in their environment. **PHYSICAL CHARACTERISTICS OF LIVING THINGS**

CHAPTER 4

HOW DO LIVING THINGS SURVIVE AND

The red-eyed tree frog is well adapted to living in a tropical rain forest. Its long legs are good for climbing trees. On the end of each toe is a sucker pad that lets the frog stick to leaves and branches. Its green back makes it hard to see among the leaves. But the frog's bold red eyes can startle a predator and give the frog a chance to flee.

Red-eyed tree frogs live in the rain forests of Central America.

CHANGE?

SCIENCE VOCABULARY

behavior (bi-HĀV-yur)

Behavior is any way that an animal interacts with its environment. (p. 144)

Hunting is a behavior that helps dragonflies survive.

instinct (IN-stinkt)

An **instinct** is an inherited behavior that an animal can do without ever learning how to do it. (p. 144)

Wasps build their nests by instinct.

learning (LUR-ning)

Learning is a change in behavior that comes about through experience. (p. 145)

This tiger cub is learning how to fish by watching its mother.

my Science Vocabulary

behavior (bi-HĀV-yur)
communication (kuh-MYŪ-ni-KĀ-shun)
habit (HAB-it)
instinct (IN-stinkt)
learning (LUR-ning)

TECHTREK
myNGconnect.com
Vocabulary Games

habit (HAB-it)

A **habit** is a behavior that is learned through practice. (p. 147)

This bison has developed the habit of not running away from snowmobiles.

communication (kuh-MYŪ-ni-KĀ-shun)

Communication is any behavior that lets animals share information. (p. 150)

Bird songs are one kind of communication in the animal world.

Physical Characteristics of Living Things

Animal Adaptations The many sharp teeth of the shark are there for a reason. They help the shark catch fish to eat. The shark's teeth make it a very good predator.

The shark's teeth are an adaptation. Adaptations are characteristics that help organisms survive or reproduce. They are inherited characteristics that are passed from parents to offspring.

Like the shark, many other predators have sharp teeth to help them kill and eat other animals.

Animals that eat plants have different kinds of teeth. For example, cows have teeth that are broad and flat. They use their teeth to grind up tough grasses, stems, and leaves. Many animals use their teeth for self-defense as well as for feeding.

The feet of animals are also adapted to different ways of life. Many animals have feet that help them find food or escape from predators. Look at the animal feet in the chart. How do the feet of each animal help it survive?

This shark is well armed for hunting its prey.

COMPARING ANIMAL FEET

SHARP TALONS
This hawk holds its prey with its long, sharp claws. The claws are called talons.

SUCTION CUPS
This foot belongs to a gecko. The bottom of the foot is covered with tiny hairs that help it grip surfaces.

STURDY HOOVES
These hooves belong to a reindeer. They are very hard and tough, which lets the reindeer run quickly over rocks and ice.

Science in a Snap! Model Deer Ears

Work with a partner. Stand four meters apart. Listen as your partner says something in a soft voice. Then cup your hands behind your ears, and listen again.

Cup your hands in front of your ears, and listen to your partner again.

A deer has ears that can move in different directions. How does this adaptation help a deer survive?

Plant Adaptations

Plants have many adaptations that help them survive in their environments. Pitcher plants have an unusual adaptation. They are able to trap insects. They then digest the insects to get nutrients. This allows the pitcher plants to grow in places where the soil has few nutrients.

How do pitcher plants trap their prey? These plants have large, pitcher-shaped leaves with sticky liquid at the bottom. Insects slide down the slippery inside walls of the pitcher and drown in the liquid. The liquid slowly breaks down the dead insects and releases nutrients that the plant needs to survive.

This "pitcher" is an adaptation that allows the plant to capture insects.

Flowers are also examples of plant adaptations. Flowers attract pollinators such as insects. One way flowers attract pollinators is with bright colors. Different colors attract different animals.

Pollinators visit the flowers and carry pollen from one plant to another. Flowering plants must be pollinated in order to make seeds. Without pollinators, many plants could not reproduce.

FLOWER COLOR AND POLLINATORS

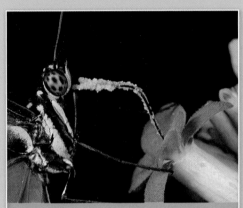
Orange flowers attract butterflies. This butterfly's long mouth part is covered with pollen.

Red flowers attract hummingbirds. Some also attract butterflies.

Yellow flowers attract bees. They may also attract other insect pollinators.

At night, white flowers are the most easily seen. They attract bats and moths that are active then.

Acquired Characteristics

Most characteristics of organisms, such as eye color, are inherited. But many characteristics can be influenced by the environment. These characteristics are called acquired characteristics. Unlike inherited characteristics, acquired characteristics cannot be passed on to offspring.

Look at the oak tree growing in the open field. Its branches are spread wide. Compare it to the oak trees growing in the forest. In the forest, trees grow upward toward the light. Their branches do not have enough room to spread out. The oak trees in the forest are the same species as the oak tree in the field. But their environment has made them have different shapes.

This lone oak tree in the middle of a field looks very different than oak trees clustered together in the forest.

Animals can also have acquired characteristics. For example, animals that do not get enough food may be small. Injuries can leave scars or cause other changes. Look at the dog in the picture. It has lost its leg.

Acquired characteristics may stay with an animal throughout its life. But acquired characteristics cannot be passed on to the next generation. Only inherited characteristics are passed from parents to their offspring.

A missing leg is an acquired characteristic that this dog has learned to live with.

Before You Move On

1. What are adaptations?
2. How do acquired characteristics differ from adaptations?
3. **Predict** A newly discovered plant has bright yellow flowers. What kind of pollinator do you predict it will have? Why?

Behaviors Help Animals Survive

Instinct and Learned Behavior The dragonfly on this page has a nickname. It is called the "pondhawk." The name fits because the pondhawk is a fierce predator. It chases and catches insects. It even preys on insects its own size.

Hunting is just one of many ways that animals behave in order to survive. Behavior is any way that an animal interacts with its environment. Besides finding food, behavior includes ways that animals protect themselves, make homes, find mates, and raise their young.

Animal behaviors may be either instinctive or learned. An instinct is a behavior that an animal can do without ever learning how to do it. Instincts are inherited characteristics. An instinct is always performed the same way. For example, dragonflies hunt by instinct. They always swoop down and capture their prey in the air. Most of the behaviors of insects are instinctive.

The green clearwing dragonfly on the right is chasing its prey—the red milkweed beetle on the left.

This Sumatran tiger is teaching her cub how to fish.

Behaviors that are not instincts must be learned. Learning is a change in behavior that comes about through experience. For example, a dog may learn to go to the back door when it wants to go outside. Learning is more important in animals with bigger brains. More intelligent animals can learn more, and more of their behavior is learned.

Young animals learn behaviors mainly from their parents. Look at the tigers in the picture. The cub is not just tagging along with its mother. It is learning how to find and catch fish by watching its mother do it.

Both instinctive and learned behaviors are important. Is one type of behavior better than the other? It depends on the situation.

Instinctive behaviors do not need to be learned, and they are always the same. However, if the environment changes and an instinctive behavior no longer works, the behavior cannot be changed.

It takes time and effort for an animal to learn a behavior. But learned behaviors are flexible. If the environment changes, a new behavior can be learned.

Protection Many behaviors help protect animals from predators. For example, opossums avoid predators by "playing dead." A frightened opossum goes limp and does not move. Predators usually leave it alone because they think it is dead. This is where the expression "playing possum" comes from.

The moth on this page has a behavior that helps scare away predators. When a predator approaches, the moth opens its wings. The two large spots on its lower wings look like the eyes of a fierce predatory bird. Spreading its wings for this purpose is an instinctive behavior.

Some animals have behaviors that help them hide from predators. Squid release a dark inky substance into the water when a predator is near. The ink blocks the predator's view of the squid, so the squid can quickly swim away.

When the moth's wings are closed, you cannot see its large spots. But when its wings are open, the spots look like an owl's eyes.

Many animals protect themselves by running away. A chipmunk darts under a rock to escape from a fox. A deer bolts through a field to escape from a chasing dog.

Other animals stand their ground and defend themselves. They may have characteristics such as sharp teeth, claws, or antlers that they can use to fight. They may also have characteristics that make them look large and fierce. Have you ever seen a cat raise its fur and arch its back when it is afraid? Reacting this way makes the cat look bigger and more dangerous.

Animals sometimes learn to stop running away from an "empty" threat. Look at the bison in the picture. It has learned that the snowmobiles are not dangerous, so there is no need to run away. This is an example of a **habit**. A habit is a behavior that is learned through practice. By staying put, the bison saves time and energy that can be put to better use, such as finding food.

A bison in Yellowstone National Park watches snowmobiles roar by.

Shelter and Raising Young Many behaviors provide shelter for animals or their young. Read about some of these behaviors in the chart.

Most animals do not take care of their young. But some animals, including birds and mammals, protect their offspring. Birds protect their eggs until they hatch. Then they feed their nestlings until they are big enough to get food for themselves.

Animals that live together in groups, such as paper wasps, often work together for the good of the group. Working together for the common good is called cooperation. Insects that live in colonies have many cooperative behaviors, such as building a nest and caring for young. These behaviors are instinctive.

Ring-tailed lemurs feed and protect their young. Lemurs live in Madagascar.

Migration Many kinds of animals migrate, or move to a different place when the seasons change. Migration is an instinctive behavior. In summer, many birds nest and raise their young in northern regions. Then they fly south for the winter. By migrating, these birds find enough food to survive.

Migrating animals don't have maps or GPS, so how do they know where to go? The position of the sun or landmarks such as rivers or mountains may help them find their way. Some animals can even sense the pull of Earth's magnetic north pole. It's as though they have a built-in compass to guide them.

ANIMAL SHELTERS

Digital Library

In Africa, male weaver birds weave grasses into complicated nests. The nests help the male birds attract females. The females lay eggs and raise young in the nests.

Pocket gophers dig burrows under the ground. They have separate rooms for sleeping and storing food. Each pocket gopher has its own burrow.

Paper wasps live in colonies. Members of a colony work together to build a nest. They use small pieces of plants and their own saliva to make the papery material.

Communication What do a hissing cat, a singing bird, and a dog that is wagging its tail have in common? All three are examples of animal communication. Communication is any behavior that lets animals share information.

Many animals communicate to attract mates and to protect themselves from predators. Animals may communicate with sounds, movements, smells, or by other means.

Look at the frilled lizard on this page. It communicates by flaring out a ruffle of skin around its neck. Can you guess what the lizard is communicating when it flares out its frill? It is saying, "I am a threat. Stay away!"

The frilled lizard is not as frightening when its frill is not flared.

Communication with sounds is important in the animal kingdom. Male birds sing to attract mates. Singing is an instinctive behavior in birds. However, young birds may have to learn their song by listening to other birds before they get it exactly right.

Many birds use calls to warn each other of predators. Mammals such as meerkats and monkeys also use warning calls. They even have different warning calls for different predators, such as snakes or birds of prey. How might this information help members of the group survive?

This bokmakierie communicates by singing and making other sounds.

Before You Move On

1. What is an instinct?
2. How does learned behavior differ from instinctive behavior?
3. **Infer** How does communication help animals learn?

NATIONAL GEOGRAPHIC

ALL HANDS ON DECK!
SAVING RIGHT WHALES

All of a sudden, a huge, dark creature shoots out of the ocean. Then it slaps the water with its flippers before disappearing again. It was a North Atlantic right whale! They aren't the largest type of whale. Even so, they are about the size of a school bus.

Right whales can dive deep into the ocean and hold their breath for up to 40 minutes. Like all whales, they are mammals, so they have to come back to the surface to breathe.

The name right whale comes from early whalers. They believed this was the "right" whale because it was easy to hunt. And hunt them they did! From the whales, they got oil for lamps and supports for umbrellas and fancy dresses. After hundreds of years of whaling, North Atlantic right whales were nearly extinct. By 1935, it was against the law to kill right whales. Fewer than 400 North Atlantic right whales are living today, making them the most endangered of all large whales.

Right whales are slow swimmers, but they are like acrobats when they jump out of the water.

North Atlantic right whales live along the East Coast of North America. During summer and fall, many gather to feed in the Gulf of Maine and the Bay of Fundy in Canada.

In early winter, female right whales migrate to the warmer waters off the coast of Georgia and Florida. There they give birth. Mothers stay close to their calves to nurse and protect them. During April, the mothers and calves swim north to join other right whales in feeding grounds in the Bay of Fundy and off the coast of Massachusetts.

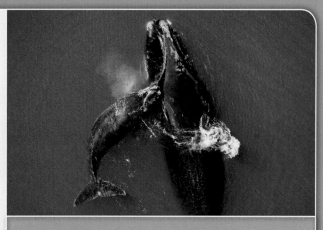

Female right whales give birth to a single calf.

Right whales use the baleen in their mouths to strain tiny animals from the ocean.

Right whales have special adaptations for feeding. Instead of teeth, they have baleen. Baleen is made of the same material as your fingernails. Hundreds of rows of baleen hang from the whales' upper jaws. The whales feed by swimming slowly with their mouths open. As water passes through the rows of baleen, tiny ocean organisms are trapped. Right whales can eat more than a ton of these tiny creatures in a single day.

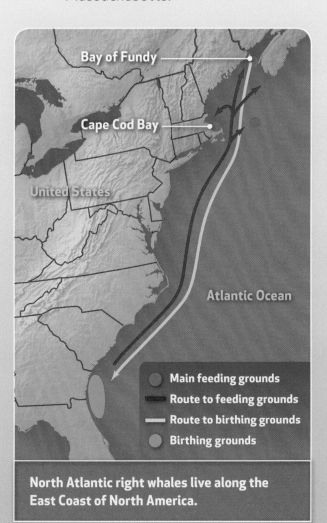

North Atlantic right whales live along the East Coast of North America.

Even though right whales are now safe from whalers, they face other serious threats. As they migrate up and down the Atlantic coast, they stay fairly close to shore. Many ships also travel through these waters. Collisions with ships are the biggest killer of right whales. Whales that do not die are often injured and scarred.

Sometimes right whales are caught in fishing ropes and nets. As they try to get free, the ropes cut through their skin. Sometimes the tangled ropes keep them from feeding, and they eventually starve.

Ocean pollution and smaller food supplies also threaten right whales. Some scientists believe that without protection, right whales could become extinct in less than 200 years.

The good news is that many people are working together to protect the right whales. Scientists, government agencies, and the fishing industry are finding ways to save whales. Preventing the deaths of just two adult females each year would give the right whales a good chance to recover.

One of the greatest threats to right whales is the ropes, nets, and traps used by fishing boats.

For 30 years, a research team from Boston's New England Aquarium has been studying right whales. These scientists have gathered almost 400,000 photographs to help them identify nearly every North Atlantic right whale. Each whale is given a number and sometimes a name. This information helps scientists track individual whales.

Tracking individual whales helps scientists learn more about the right whale population so they can plan for its recovery.

Every sighting of a right whale is sent to a warning system that lets ships know whales are nearby. U.S. and Canadian officials have lowered ship speeds and changed shipping routes where right whales are found.

New fishing rules are helping protect the whales. Fishing ships cannot use certain kinds of gear that are harmful to whales. Researchers are working on fishing lines that will break if whales get caught in them.

There are also rescue teams that work to free whales that are tangled up in fishing lines.

Are these measures helping the whales? Since 2001, the number of right whales has been increasing. In 2009, a record of 39 calves were born. And there have been far fewer deaths caused by ships and fishing gear. These are encouraging signs of recovery. But most scientists say the whales are not out of danger yet.

The white scars on its tail help scientists identify this right whale.

Life Cycle Adaptations

Plant Life Cycles The wildflowers on this page are growing in a desert. Does that surprise you? These plants have adaptations that help them survive in a dry environment. One adaptation is their ability to go through their entire life cycle very quickly after it rains. In just a few weeks, seeds germinate, stems and leaves grow, flowers bloom, and seeds form.

The plants spend the rest of the time as seeds. Their seeds can survive dry conditions for months, or even years, until it rains again.

In many climates, the weather is too cold for plants to grow during the winter. Plants can grow only during spring, summer, and fall. Many flowering plants die when the weather turns cold. When spring returns, new plants must grow from seeds again. Plants with this type of life cycle may produce very large numbers of seeds each year.

For most of the year, the weather is dry and few plants bloom.

When rain comes to Picacho Peak State Park in Arizona, the land is covered with wildflowers.

Other kinds of plants live for many years. These plants may go dormant each winter and lose their leaves or even die back to the ground. In spring, they start growing again. Some plants do not lose their leaves during the winter. These plants are called evergreens. Pine trees are an example.

Many plants survive winter or a dry season by storing food in underground bulbs. Plants that form bulbs include hyacinths and tulips. The pictures below show the hyacinth's yearly growing cycle.

YEARLY CYCLE OF A **HYACINTH**

A hyacinth is a plant that grows from a bulb and blooms in the spring.

The plant survives dry and cold weather as a bulb underground.

When the weather begins to get warm and there is enough rain, the plant sends up flowers.

The leaves continue to grow and make food.

Food moves down into the bulb and the leaves die.

Animal Life Cycles The life cycles of animals help them survive in their environment. Insects have some of the most interesting life cycles. You can see the life cycle of the monarch butterfly on the next page. Notice that a monarch has four stages: egg, larva, pupa, and adult.

In each stage of its life cycle, an insect has different needs. For example, larvae need a great deal of energy to grow and change into adults. They get the energy they need by eating constantly, day and night. Many kinds of insect larvae feed on leaves. These insects must go through their larval stage at a time of year when there are plenty of leaves. In most of the United States, this is during the spring or summer.

Insects that live in places with cold winters must have adaptations that protect them from freezing. These adaptations vary, depending on the stage at which insects go through the cold season. Many insects spend the winter as eggs. The eggs do not hatch until the weather is warm enough for the larvae to live.

Insects that spend the winter as pupae are protected inside a chrysalis or cocoon. Some adult insects find a protected place and spend the winter in a sleep-like state, similar to the hibernation of other animals. Insects that live in colonies crowd together in their nest to share body heat.

A few adult insects migrate to a warmer climate. For example, monarch butterflies migrate from Canada and the northern United States to spend the winter in Mexico.

LIFE CYCLE OF A MONARCH

Monarch butterflies go through a life cycle called complete metamorphosis.

LARVA Butterfly larvae are called caterpillars. Each caterpillar eats and grows larger.

PUPA The caterpillar forms a protective chrysalis. Inside the chrysalis, the pupa goes through many changes.

EGGS Monarchs lay their eggs on milkweed plants.

ADULT The adult has wings. It is ready to find a mate and reproduce.

Before You Move On

1. List three adaptations that protect adult insects from the cold.
2. Compare the life cycle of a plant that lives for just one year with the life cycle of a plant that lives for many years.
3. **Apply** Many mammals such as groundhogs hibernate, or sleep deeply during the winter. How does this behavior help the groundhogs survive?

When Environments Change

Differences Among Individuals These wild horses all belong to the same species. In many ways they are just alike. For example, they all have pointed ears, two eyes, and a mane. But the horses are not alike in every way. Each horse has inherited characteristics that make it slightly different from the others. What differences do you see?

The most obvious difference is the color of their hair. But the horses also differ in many less obvious ways, such as their strength and speed.

The individual members of all species differ in many ways. Some of these differences affect their ability to survive. For example, wild horses that can run faster may be better able to escape from predators.

Although all the horses in this herd are similar, individual horses differ from each other.

Plants as well as animals show variation in their characteristics. Compare the two bush lilies shown here. Both plants belong to the same species, but their flowers are different colors. Flower color can affect a plant's ability to reproduce, because colorful flowers attract pollinators such as insects.

Yellow bush lilies and red bush lilies are found in different places. Yellow bush lilies are found where bees are common pollinators. What pollinators do you think are common where red bush lilies are found?

Red bush lilies attract hummingbirds as pollinators.

Yellow bush lilies attract bees as pollinators.

Changes in Populations

You can't see them in the big picture, but many animals live in this desert. Rock pocket mice live in small burrows near the rocks.

Most rock pocket mice have brown fur, but some have black fur. In most places, brown mice are more common than black mice. But in places with black rocks, black mice are more common. Can you explain why?

On black rocks, it is hard for predators to see black mice. Their black fur blends in with the rocks. Since they are harder to catch, black mice are more likely to survive than brown mice. Therefore black mice live longer and have more offspring. Over time, mice with black fur become more common.

When some individuals of a population are better adapted than others, their characteristics can be passed on to the next generation. Over time, their characteristics become more common in the population.

Brown rock pocket mice blend in with light-colored rocks. This helps them hide from predators.

On dark-colored rocks, it is harder to see black rock pocket mice.

When environments change, differences among individuals may allow some organisms to survive while others die. Scientists observed this happening on the Galápagos Islands, which are in the Pacific Ocean.

Some Galápagos finches eat seeds, which they crack open with their beaks. Birds with bigger beaks can crack open bigger seeds. During several dry years, seeds were scarce. Then many birds with smaller beaks could not find enough small seeds to survive.

Birds with bigger beaks could crack and eat more seeds. Most of these birds survived and reproduced. In just a few generations, most of the finches had larger beaks.

Some organisms survive change by moving to a new location. Recently, summers in the Arizona desert have become hotter and drier. Some plants cannot live where it is so hot and dry. But seedlings of these plants now grow higher up the mountainside. How does this help? Higher on the mountains the weather is cooler and wetter, so the plants can survive.

Finches with small beaks can crack open only small seeds.

Finches with large beaks can crack open both large and small seeds.

Extinction When environments change, organisms may not have the characteristics needed to survive in the new habitat. Instead, they may go extinct. This happened to the dusky seaside sparrow shown here.

Dusky seaside sparrows used to live in salt marshes in Florida. They began dying off in the 1940s, when people sprayed poison on marshes to kill mosquitoes. The birds were poisoned when they ate the insects. Then people started draining the marshes. This destroyed most of the birds' habitat. The species was declared extinct in 1990.

Many species have recently gone extinct, and many more species are at risk of extinction in the future. Extinction is nothing new. As many as 99 percent of all species that ever lived on Earth have gone extinct. But recently the rate of extinction has been increasing. Pollution, the introduction of new species and diseases, and habitat loss are the main causes of modern extinctions.

The dusky seaside sparrow used to live in salt marshes on Merritt Island in Florida before it went extinct.

Merritt Island National Wildlife Refuge, Florida

Animal extinctions get more attention, but many plants have also gone extinct. One example is the Franklin tree, which was named for Benjamin Franklin. It still grows in gardens but no longer grows in the wild. The rise of cotton farming in Georgia, where the tree used to grow wild, may explain why. A fungus that grows on cotton plants also attacks and kills the tree.

Humans have protected other species from going extinct. The bald eagle almost died out in the 1960s. The same poison that killed the dusky seaside sparrow was killing eagles. But the poison was banned, and the bald eagle started to make a comeback. By the 1990s, it was no longer in danger of extinction.

Sometimes organisms thought to be extinct have later been found alive. The coelacanth fish was thought to have gone extinct 65 million years ago. But in 1938, scientists were surprised when a living coelacanth was found. Since then, several more have been discovered.

The coelacanth fish was once thought to be extinct.

Evidence of Life in the Past

How do scientists know about organisms that are now extinct? The answer is fossils. Preserved remains of dead organisms are sometimes found in rocks. Fossils show that many different species lived in the past. By studying fossils, scientists can learn about extinct organisms and their environments. For example, fossils of fish found in places that are now dry land show that these places were once covered by lakes or seas.

Plants as well as animals may become fossils.

As many as 85 percent of all species went extinct 65 million years ago. That's when all the dinosaurs died out. This is one of several mass extinctions that have occurred during Earth's history. There are several possible causes of mass extinctions.

One is a huge asteroid striking Earth. This could have caused clouds of dust and gas that blocked sunlight from reaching Earth's surface. If the asteroid crashed into the ocean, it would have caused a tsunami, or very high ocean wave. The wave would have flooded land near the coast. Many volcanoes erupting around the same time is another possible cause of mass extinctions.

These scientists are studying fossils of a giant dinosaur called Titanosaurus.

A few organisms have managed to survive environmental change without going extinct. They are called living fossils. Horseshoe crabs are living fossils. Compare the pictures of the living and the fossil horseshoe crabs. How are they alike?

Other examples of living fossils are sharks, cockroaches, and gingko trees. All of them have changed very little for hundreds of millions of years. How do you think these living fossils have survived unchanged for so long?

You can see from the fossil on the left that horseshoe crabs have changed little over millions of years.

Before You Move On

1. What are some ways that individual plants and animals of the same species differ?
2. Explain how the characteristics of a population can change over time.
3. **Predict** One species of plant lives only on a small volcanic island. The volcano erupts and covers part of the island with ash. How might that change in the environment affect the survival of the species?

Conclusion

Living things have many different adaptations that help them survive. Most adaptations are inherited. Animals depend on their behaviors to survive. Behaviors may be instinctive or learned. Some animals have behaviors that let them communicate information to other animals. The life cycles of plants and animals are also adaptations that aid in survival. When environments change, differences among individual organisms allow some to survive while others die or move to other locations.

Big Idea Organisms have many different kinds of adaptations that let them survive and reproduce in their environment.

Adaptations include physical characteristics and behaviors.

Vocabulary Review

Match each of the following terms with the correct definition.

A. behavior
B. communication
C. habit
D. instinct
E. learning

1. Behavior that is learned through practice
2. Any behavior that lets animals share information
3. Change in behavior that comes about through experience
4. Any way that an animal interacts with its environment
5. Inherited behavior that an animal can do without ever learning how to do it

Big Idea Review

1. **List** What are three ways that animals use communication?

2. **Describe** Describe an adaptation that helps protect some insects from the cold.

3. **Relate** How do the colors of flowers relate to pollination?

4. **Cause and Effect** How does migration help animals survive?

5. **Infer** Assume that all the members of a species are just alike. How do you think this would affect their ability to survive if their environment changes?

6. **Make Judgments** What responsibility do you think human beings have to protect other living things from extinction?

Write About Animal Behavior

Explain Look at the weaver bird in the picture. Describe what it is doing. Explain how its behavior helps it reproduce.

NATIONAL GEOGRAPHIC

CHAPTER 4 — LIFE SCIENCE EXPERT: AQUATIC ECOLOGIST

Zeb Hogan

Dr. Zeb Hogan leads the National Geographic Megafishes Project, which works to study and protect Earth's largest freshwater fishes and the rivers where they live. Zeb is an aquatic ecologist at the University of Nevada, Reno. He has studied fishes on six continents—North America, South America, Europe, Africa, Asia, and Australia.

Zeb Hogan with a taimen in Mongolia

What does an aquatic ecologist do?

Aquatic ecologists study the interplay between aquatic organisms, such as fish, and their environment. My specialty is megafish. Megafish are freshwater fish that grow more than 6 feet (1.8 m) long or weigh more than 200 pounds (91 kg). There are about two dozen species of megafish in the world.

What is a typical day for you in the field?

In the field, I spend most of my time on or near the water. In Southeast Asia, I rely on fishers to help me gather information. We ask them about the number and kinds of fish they are catching. If they catch an endangered fish, we help them tag it and release it back into the water.

I also work with National Geographic to produce news stories and shows for television. This gives me a chance to share my experiences with many people.

Zeb shows off a giant stingray in Cambodia.

Can you judge the size of this taimen by looking at Zeb's outstretched arms?

What has been your greatest accomplishment so far?

I'm most proud of my work with the Mekong giant catfish. I was part of a team that helped to list the catfish as Critically Endangered. Since that listing, we've convinced some fishers to stop fishing for the catfish. We are now working with other scientists and government officials to find the best ways to protect the species.

Why is your work important?

Some megafish, such as sturgeon, have lived on Earth for over 100 million years. I feel strongly that these fish and the rivers where they live need our help. The conservation of these unique species is important. They are indicators of the health of freshwater ecosystems.

NATIONAL GEOGRAPHIC

BECOME AN EXPERT

The Amazing World of Ants

Ants All Around Us Imagine trying to pick up and carry a car—with your jaws! Humans cannot lift things that are so much heavier than themselves, but ants can. Ants can pick up things that are many times their own weight and carry them long distances. Ants may be small, but they are mighty.

There are many different kinds of ants—more than 10,000 species. Ants live just about everywhere on land except near the poles. In some places, one out of every two insects is an ant. Each ant is very small, but there are huge numbers of them. Together, they make up 20 percent of the total mass of land animals. In fact, ants have a greater mass than all land vertebrates combined!

The key to ants' success is their social behavior. All ants live in groups called colonies. Ant colonies range in size from dozens to millions of ants.

These ants are holding onto each other to form a living ladder.

behavior
Behavior is any way that an animal interacts with its environment.

TECHTREK
myNGconnect.com

Student eEdition

Digital Library

Through **instinct**, members of a colony cooperate in everything they do. By working together, they can do many things that a single ant could never do alone. The ant ladder in the picture is a good example. Other ants can scramble up the ladder to reach a high branch that a lone ant would not be able to reach.

Most ants will eat just about anything, but some prefer certain foods, such as insects or seeds.

Many ants like sweet liquids, such as plant sap or honeydew, a sweet liquid secreted by insects called aphids. Some ants even "herd" aphids to have a steady supply of honeydew. Only a few ant species, such as leafcutter ants, eat a single kind of food. Leafcutter ants feed only on fungus. They grow the fungus on beds of chewed leaves in their nest.

Leafcutter ants carry leaves back to their nest.

instinct
An **instinct** is an inherited behavior that an animal can do without ever learning how to do it.

173

BECOME AN EXPERT

Ant Castes

Like many other insects, ants undergo complete metamorphosis. They go through four life stages: egg → larva → pupa → adult.

Each adult belongs to one of three different classes, or castes: queen, drone, or worker. Each caste looks different from the others and has a different job in the colony.

The queen is the head of the colony. Her only job is to lay eggs. She may lay up to 1,000 eggs a day, day after day for many years. She needs to mate just once to have all of her eggs fertilized. A queen has wings until she mates. After that, she breaks off her wings and stays in the nest.

Drones are the only males in the colony. Drones do not do any work. After they mate with a queen, they die.

A queen fire ant with pupae and workers

In many ant species, there are different types of workers. For example, some workers may have adaptations to be soldiers. The job of soldiers is to defend the nest from predators. How is the soldier ant below well suited for this job?

In a few ant species, some workers are even more specialized. For example, in honeypot ants, some workers have adaptations for storing food inside their body (see picture).

An ant's caste depends mainly on the food it gets as a larva. Larvae that get the food with the most nutrients develop into queens. Others develop into drones or workers. Because caste is affected by the environment in this way, an ant's caste is not fixed when it hatches. This makes ant colonies flexible. They can adjust the numbers in each caste to meet changing needs. This improves the colony's chances of surviving.

This soldier ant has large, powerful jaws. It uses its jaws to bite predators.

These honeypot ants eat and store sweet liquids inside their body until they swell to the size of a marble. They spit up and share some of the liquid when other workers tap them with their antennae.

BECOME AN EXPERT

Ant Homes

Once a year, new adult drones and queens emerge from the pupa stage. They soon fly into the air and mate. Then the drones die, while the queens start new colonies. Each queen digs a small nest and lays her first batch of eggs. As the eggs hatch and complete the life cycle, worker ants emerge. The workers expand the nest and take over all the other work of the colony.

Most ants dig underground nests with many chambers, or rooms, connected by tunnels. Some chambers are used for nurseries, where eggs, larvae, or pupae are cared for.

Other chambers are used for storing food or for egg laying by the queen. In cold climates, there may be chambers deep below the surface where ants can go in the winter to stay warm. Ant nests may be very deep. Some reach more than 10 meters (35 feet) below the surface.

An ant nest may have hundreds of chambers and many meters of tunnels.

REST AREA
Worker ants take a break between jobs.

SEED STORAGE
Workers stash seeds here to eat later.

NEW ROOM
As a colony grows, workers add rooms.

Not all ants dig underground nests. Some make nests in trees or dead logs. A few species do not build nests at all. Army ants, for example, travel all day, eating just about everything in their path. Each night, they cluster together in a different tree.

An ant colony may last for many years and grow to have millions of members. Queens can live for up to 30 years. Most colonies last until the queen dies.

Without a queen, the colony will not last long. The worker ants die off, and there are no new workers to replace them.

A worker ant is a wingless female that cannot reproduce. Workers do all of the work of the colony, including finding food.

NURSERIES
Young ants live here after they are born.

TUNNEL
Passageways link chambers together

QUEEN'S CHAMBER
Here the queen lives and lays eggs.

WINTER QUARTERS
Ants move to the deepest chambers during cold weather.

BECOME AN EXPERT

Ant Communication and Learning

Ants could not live and work together in such large numbers without forms of communication. The main way ants communicate is with chemicals. In fact, they have the most complex chemical communication in the animal kingdom. They give off chemicals, called pheromones, from special glands. They smell the pheromones of other ants with their antennae.

Different ant pheromones mean different things. For example, some pheromones mean "danger." Other pheromones show which colony an ant belongs to. Still others are used to mark trails. Ants may travel up to 200 meters (700 feet) from their nest looking for food, so marking trails is important. It helps them find their way back to the nest, and it shows other ants where to find food.

These ants are using their antennae to identify each other as nestmates.

communication
Communication is any behavior that lets animals share information.

Ants also communicate with sound, touch, and taste. They make sounds by rubbing their mouth parts on their body. They don't have ears, but they can sense sounds through their feet and antennae. Ants also use their antennae to feel things, especially in their dark underground nests. They use their antennae to taste things as well. No wonder ants always seem to be moving their antennae! It's not a **habit**. It's their instinctive way of sensing the world.

Most ant behaviors are instinctive, but **learning** also takes place in ants. For example, older workers have been seen teaching younger workers where to find food. The teachers lead the students along a trail to a food source. They slow down whenever the students lag behind, until the students catch up.

Unlike most other insects, ants are able to bend their antennae. This makes their antennae more useful for sensing the environment.

habit
A **habit** is a behavior that is learned through practice.

learning
Learning is a change in behavior that comes about through experience.

BECOME AN EXPERT

CHAPTER 4: SHARE AND COMPARE

Turn and Talk How does living in colonies help ants survive? Form a complete answer to this question together with a partner.

Read Select two pages in this section. Practice reading the pages. Then read them aloud to a partner. Talk about why the pages are interesting.

Write Write a conclusion that tells the important ideas you learned about the behavior of ants. State what you think is the Big Idea of this section. Share what you wrote with a classmate. Compare your conclusions.

Draw Imagine what it would be like to be a worker ant, living in a colony with many other ants. Decide on the type of worker you would be and draw a picture of an activity you might do. Combine your drawing with those of your classmates to make an ant colony mural.

LIFE SCIENCE

After reading Chapter 5, you will be able to:

- Identify the purpose of organ systems in humans. **ORGAN SYSTEMS**
- Explain how organ systems work together to perform the body's activities.
  SKELETAL AND MUSCULAR SYSTEMS, CIRCULATORY AND RESPIRATORY SYSTEMS, DIGESTIVE AND EXCRETORY SYSTEMS, THE NERVOUS SYSTEM
- Compare organ systems in different kinds of vertebrates. **COMPARING ORGAN SYSTEMS**
- **Science in a Snap!** Compare cardiac muscle with skeletal muscle.
  CIRCULATORY AND RESPIRATORY SYSTEMS

CHAPTER 5

HOW DO BODY SYSTEMS WORK

Even when you're asleep, your body is working. So think about how busy the different parts of this person's body are! What parts seem to be working the hardest? Compare her action to what you're doing right now. What parts of your body are working the hardest?

STEMS TOGETHER?

Student eEdition | Vocabulary Games | Digital Library | Enrichment Activities

The ballerina slips to the floor and holds her arms in this pose. How does her body do this?

SCIENCE VOCABULARY

organ system
(OR-gun SI-stem)

An **organ system** is a group of organs that works together to do a specific job in the body. (p. 187)

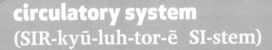

Organ systems enable you to run, jump, and play.

circulatory system
(SIR-kyū-luh-tor-ē SI-stem)

The **circulatory system** is a group of organs that carries blood and oxygen to the body and wastes away. (p. 192)

The boy's circulatory system works hard when he runs to protect the goal.

respiratory system
(RES-pur-uh-tor-ē SI-stem)

The **respiratory system** is a group of organs that takes oxygen into the body and removes wastes. (p. 196)

The girl uses organs in her respiratory system to blow up the balloon.

my Science Vocabulary

circulatory system (SIR-kyū-luh-tor-ē SI-stem)

digestive system (dī-JES-tiv SI-stem)

excretory system (EKS-kre-tor-ē SI-stem)

nervous system (NER-vus SI-stem)

organ system (OR-gun SI-stem)

respiratory system (RES-pur-uh-tor-ē SI-stem)

TECHTREK
myNGconnect.com
Vocabulary Games

digestive system
(dī-JES-tiv SI-stem)

The **digestive system** is a group of organs that breaks down food into nutrients the body can use. (p. 198)

> The orange slices will move through the girls' digestive systems.

excretory system
(EKS-kre-tor-ē SI-stem)

The **excretory system** is a group of organs that removes wastes from the blood that the body cannot use. (p. 200)

> These children's excretory systems help keep them healthy.

nervous system
(NER-vus SI-stem)

The **nervous system** is a group of organs that takes in information from the surroundings and tells the body how to respond. (p. 202)

> Your brain is part of your nervous system.

185

Organ Systems

The human body is made up of dozens of organs. These structures carry out specific jobs as you live and grow. Your skin is your largest organ. It covers your entire body. Unlike most other organs that stay about the same size throughout your life, the body adds more skin as you grow larger.

Your skin has certain jobs. Besides protecting everything inside your body, it helps you cool down when you get hot. Also, skin helps your body get rid of wastes that are produced when you perspire. The skin includes parts that grow hair and that enable you to feel hot and cold.

These runners feel the cool breeze because of structures in their skin!

You can't see most of your organs. They're tucked away inside your body. Your brain is covered by the bones of your skull. Other organs, such as your stomach, are protected by thick layers of muscles. The photo below was made by a special machine that uses magnets and radio waves to make images. Images such as these help determine whether an organ is working correctly.

Your organs also work together in teams. A group of organs that works together is an organ system. For example, your heart continually pumps your blood, but that blood could not move throughout your body without your blood vessels. Your heart, blood, and blood vessels work together as a team to move oxygen, food, and wastes throughout your body.

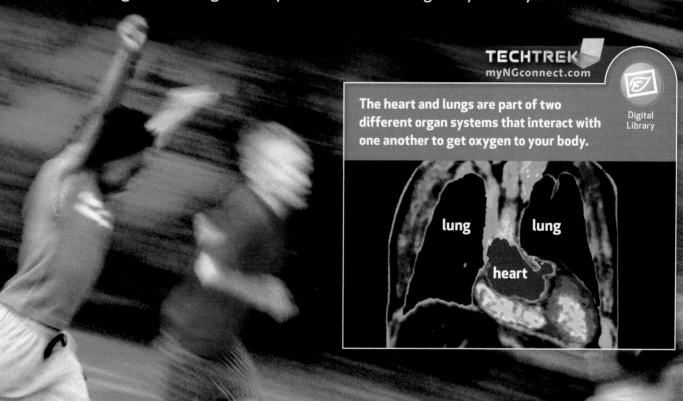

The heart and lungs are part of two different organ systems that interact with one another to get oxygen to your body.

Before You Move On

1. What are organs?
2. Explain the functions of your skin.
3. **Apply** Make a comparison between the team members of a choir and the team members in an organ system.

Skeletal and Muscular Systems

Skeleton Think about a tent. Its shape is determined by the frame inside it. Your body is somewhat like that. Inside your skin you have a framework of bones. Bones are organs.

This framework of bones is called the skeletal system. In addition to providing a support structure, the skeletal system does other jobs, too. It protects other organs, helps make blood, and stores minerals that your body needs.

You might think that bones are solid organs. But most are not. If you looked inside a bone such as the one in the upper leg, you would see it has a soft material in the center and lots of holes like a sponge near the ends. These different parts of a bone help it do its jobs.

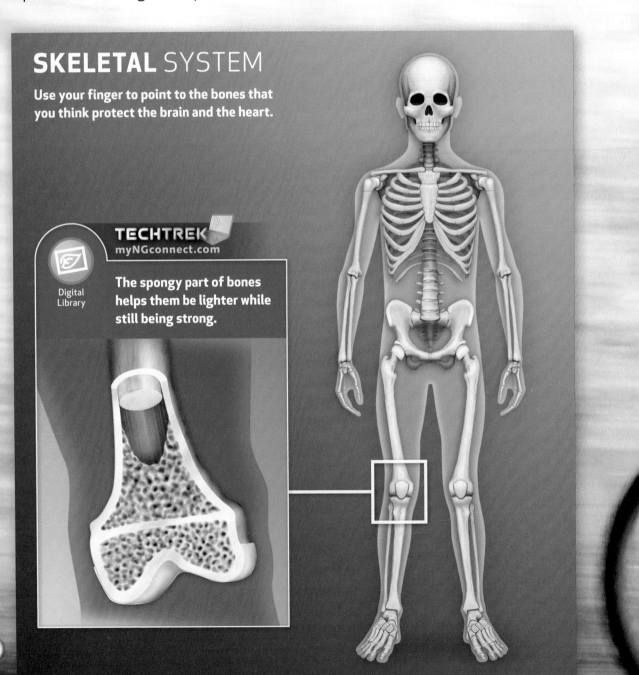

SKELETAL SYSTEM

Use your finger to point to the bones that you think protect the brain and the heart.

TECHTREK
myNGconnect.com

Digital Library

The spongy part of bones helps them be lighter while still being strong.

An adult human body has 206 bones of different sizes and shapes. Bones in your arms and legs are long and slender. Those in your hands and feet are short and stubby. Your skull is made of flat bones. The bones that make up your spine are rounded and fit together like a stack of puzzle pieces.

Bones come together at joints. Most of your joints are movable. Some joints, such as those in elbows and knees, bend like door hinges. Joints between chest bones allow slight back-and-forth movement, helping you breathe. The joints between the bones in your skull do not move at all. How do you think this is helpful?

The bony kneecap sits over the joint where the upper and lower bones of the leg come together.

Muscles The joints where bones connect determine where your body can bend, but bones cannot move on their own. Your muscular system does that job. Your muscular system is a group of organs that causes movement in the body. When any part of your body moves, inside or out, muscles are at work. Muscles contract, or pull. When muscles relax, they stretch out.

Three kinds of muscles make all the movements in your body. Skeletal muscles are connected to bones. When a skeletal muscle contracts, it pulls on a bone and causes movement at a joint. When you flex your arm or pick up something heavy, you can see some of your arm muscles bulge. Movement of your skeletal muscles is voluntary—that means you decide to move them.

This boy is using skeletal muscles to stand straight, hold his skis over his head, and smile!

But not all of your muscles are connected to bones. Organs such as the stomach, bladder, and blood vessels contain smooth muscle tissue. Smooth muscles are involuntary. That's good because it means you don't have to think about having your stomach mix up your food!

Your heart is the only organ made of cardiac muscle, which is also involuntary. It contracts and relaxes, pumping blood throughout your body your entire life. It only rests between beats!

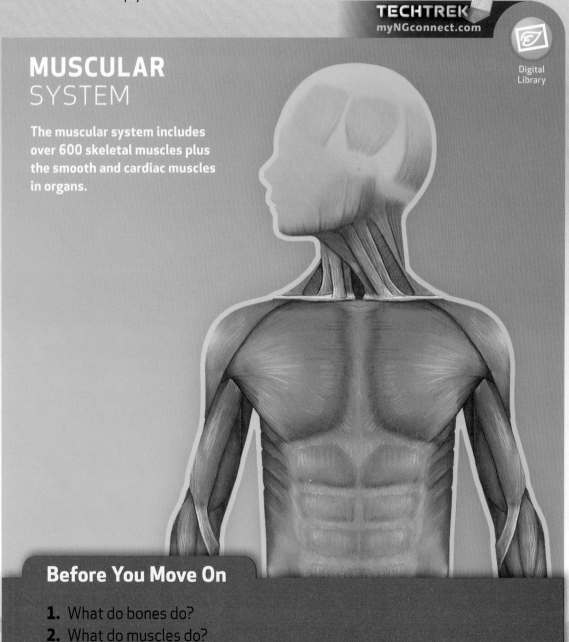

MUSCULAR SYSTEM

The muscular system includes over 600 skeletal muscles plus the smooth and cardiac muscles in organs.

Before You Move On

1. What do bones do?
2. What do muscles do?
3. **Generalize** Pick up a pencil. Explain how the skeletal system and muscular system work together in this action.

Circulatory and Respiratory Systems

Moving Blood Adults have about 5 liters (about 10 pints) of blood moving through their bodies. What does it do? Where does it go? How does it keep moving?

The circulatory system includes the blood, blood vessels, and the heart. Blood delivers nutrients and oxygen to all parts of the body and carries away wastes. Blood moves through a network of vessels made of smooth muscle. Blood circulates in blood vessels throughout your body all of the time, even while you sleep.

An artery carries blood away from the heart. Arteries have muscle in their walls, which helps push the blood throughout the body.

Arteries connect to capillaries where the exchange of materials between the body and the blood occurs. Capillaries are the tiniest blood vessels. Capillary walls are so thin that oxygen moves out of the blood, through the wall of the capillary, and into the body. Waste products move from the body back into the blood.

Capillaries lead to veins. These blood vessels carry blood back to the heart. Then the blood circulates again.

The goalie's blood is moving quickly, carrying food and oxygen to all his muscles.

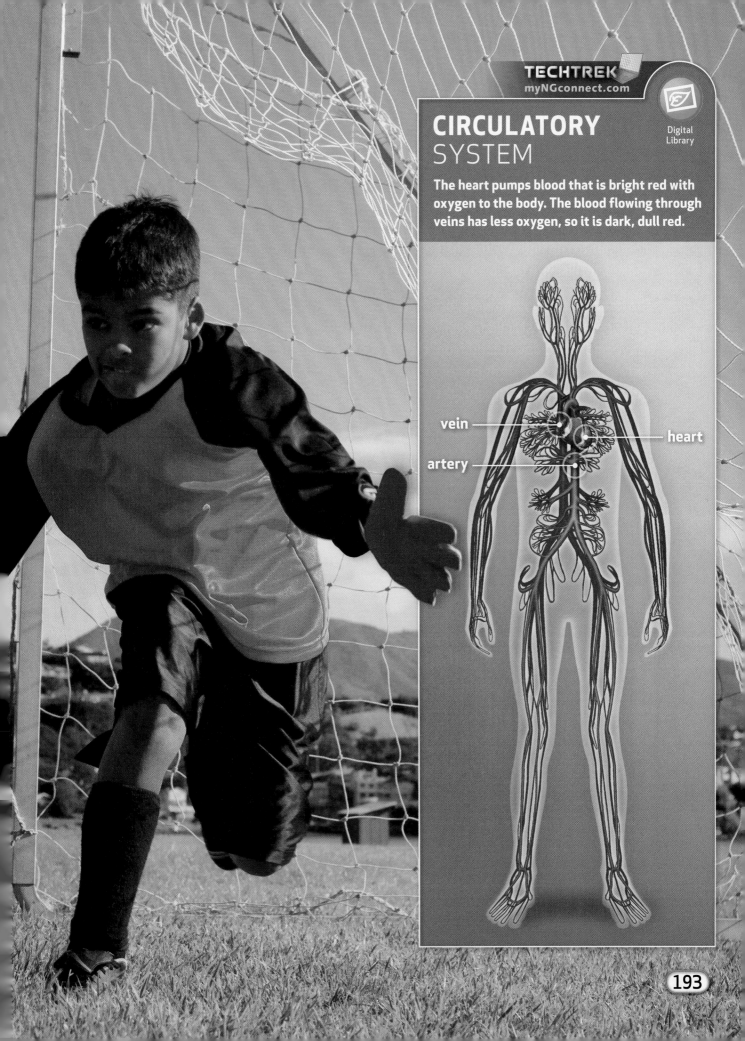

CIRCULATORY SYSTEM

TECHTREK
myNGconnect.com
Digital Library

The heart pumps blood that is bright red with oxygen to the body. The blood flowing through veins has less oxygen, so it is dark, dull red.

- vein
- heart
- artery

193

The Heart How does your blood keep moving through your blood vessels? It is pumped, nonstop, by your heart. The cardiac muscle contracts and relaxes, over and over. When the heart contracts, it pushes blood out to the body. When the heart relaxes, it fills with blood again.

The heart has four chambers, or sections, that are stacked, two above and two below. Blood from the body enters the top right chamber and then flows to the lower right chamber. This chamber pumps the blood to the lungs. Then the blood from the lungs enters the top left chamber and flows to the lower left chamber. This chamber pumps blood out to the rest of the body. That takes a big push! The lower left chamber of your heart has the thickest muscle because it pumps so hard.

Your heart pumps faster or slower depending on what activity you are doing. For example, it pumps slower when you are sleeping. It pumps faster when you are running.

Science in a Snap! Feel the Beat

Lightly place your first two fingers of one hand over the inside of the wrist on your other arm. Move your fingers around slowly. Stop when you feel a throbbing. The throbbing is your pulse.

Count the number of pulses in 10 seconds. Multiply that number by 6. This is the approximate number of times your heart beats in one minute. Then try to squeeze a tennis ball as hard as you can for as long as you can. Observe how your arm feels.

The force your heart uses to pump blood is about the same as the force you use to squeeze a tennis ball. What can you infer about differences between your heart muscle and the muscles you use to squeeze the ball?

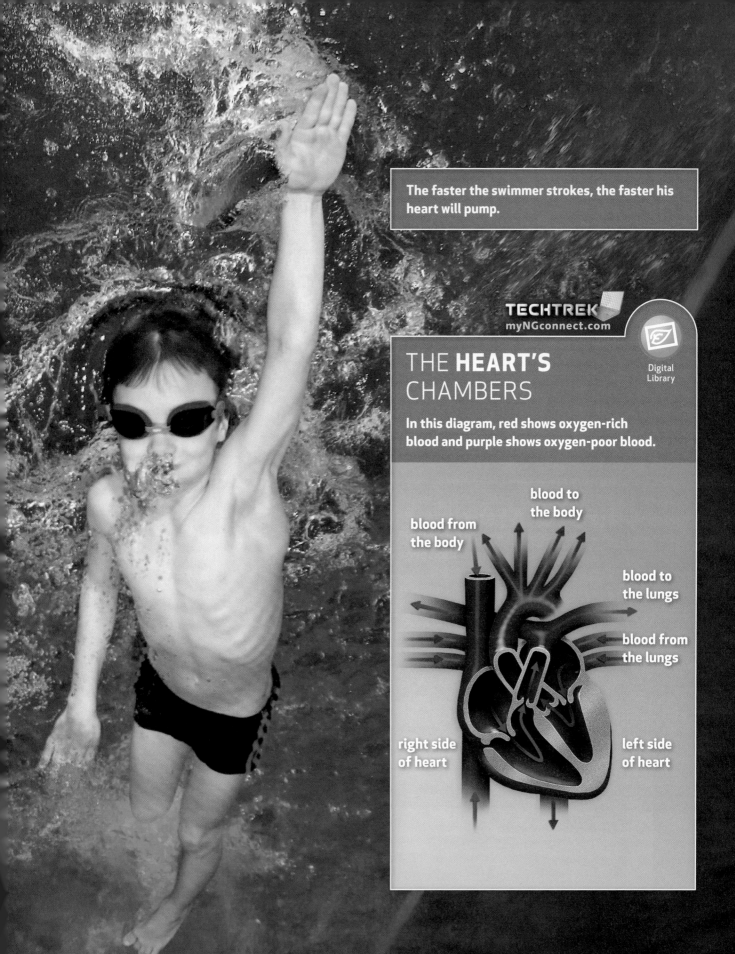

The faster the swimmer strokes, the faster his heart will pump.

THE **HEART'S** CHAMBERS

In this diagram, red shows oxygen-rich blood and purple shows oxygen-poor blood.

blood from the body

blood to the body

blood to the lungs

blood from the lungs

right side of heart

left side of heart

Getting Oxygen Organs in your respiratory system enable your blood to get the oxygen from the air that your body needs. When you inhale, or breathe in, you take oxygen into your body. Your body uses that oxygen and produces carbon dioxide as waste. When you exhale, or breathe out, you push out the carbon dioxide waste.

Your diaphragm is a sheet of muscle beneath your lungs that helps you inhale and exhale. When the diaphragm contracts, it moves downward. Your ribs move out, and your chest area gets a little bigger. Your lungs fill with air. When the diaphragm expands, it relaxes and moves back up. Your ribs move back in, and your chest area gets a little smaller. You exhale, and your lungs empty.

Just as your involuntary heartbeat continues night and day, your breathing continues all of the time. You can control your breathing for short periods, but when you stop thinking about it, your breathing continues as an involuntary function, even when you are asleep.

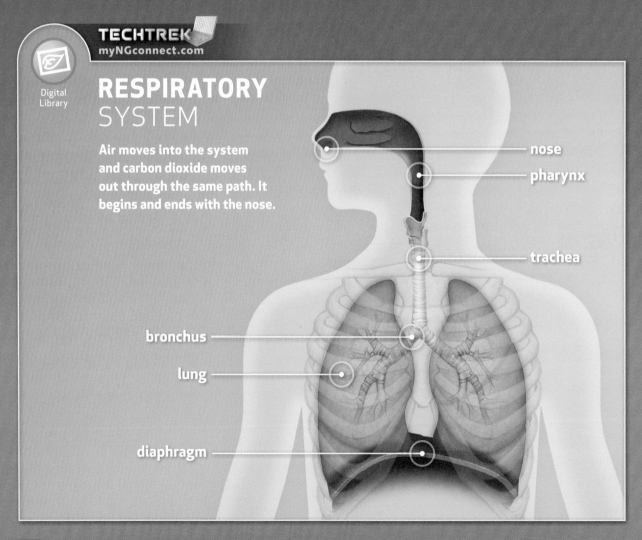

TECHTREK
myNGconnect.com

Digital Library

RESPIRATORY SYSTEM

Air moves into the system and carbon dioxide moves out through the same path. It begins and ends with the nose.

- nose
- pharynx
- trachea
- bronchus
- lung
- diaphragm

You breathe air into your lungs, which are the main organs of the respiratory system. They are like inflatable bags full of tubes that branch out like tiny trees. These tubes are the places where blood and air meet and exchange gases.

You breathe harder when you run because your muscles are working harder. They need more oxygen to do their job. When this happens, your heart beats faster to move more blood through your lungs. Your lungs must bring in oxygen and get rid of carbon dioxide more quickly.

This girl's balloon contains more carbon dioxide than the air around it.

Before You Move On

1. How does blood move through the body?
2. Explain how the heart and lungs work together.
3. **Generalize** Think about what the word *transportation* means. Explain why the circulatory system often is called your transport system.

Digestive and Excretory Systems

Getting Food Are you hungry? If so, your body is telling you it needs food. But your body cannot use the food you eat until it is digested. The organs in your digestive system break food down into nutrients the body can use.

Digesting food begins when you chew it. Chewing breaks food into smaller pieces and mixes the food with saliva. This is just one of the juices your body makes that helps in digestion. Saliva begins to digest some foods. When you swallow, the crushed food travels through the esophagus to your stomach.

The stomach is a baglike organ where food mixes with water and other juices your body makes. The muscles in the stomach cause it to churn the food so it is mixed well with the juices. The soupy mixture then moves into the small intestine. This organ is a very long, winding tube. Most digestion happens here.

Two organs that help digestion are the pancreas and the liver. Both organs make certain juices that flow through the small intestine. The pancreas makes juices that help digest proteins and starches. The pancreas also helps keep the right amount of sugar in your blood. The liver produces bile that helps digest fatty foods.

As food moves through the small intestine, the nutrients in food are absorbed for the body to use. Anything that cannot be digested moves into the large intestine. Here, water is absorbed into the blood. The waste material stays in the large intestine until it leaves the body.

DIGESTIVE SYSTEM

A hamburger takes about 24 hours for the organs to digest it.

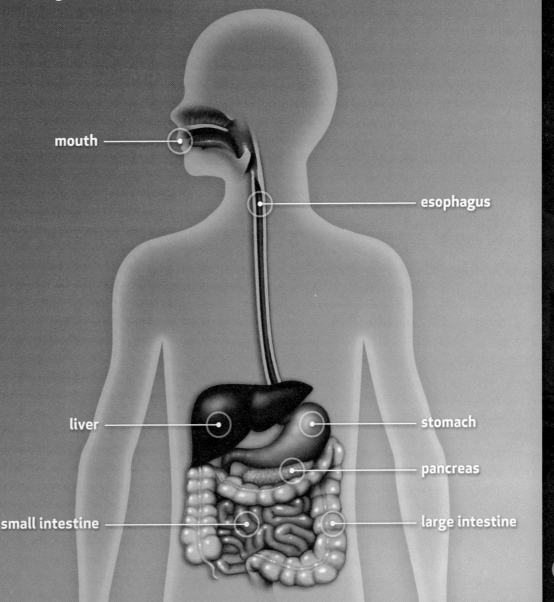

Removing Wastes Thinking, working, playing, growing, sleeping, or doing anything requires energy. You might eat a peanut butter sandwich to get energy to do something. As your body uses the nutrients from the peanut butter and bread, the body makes other materials your body cannot use. The body has to get rid of these waste materials. The **excretory system** gets rid of materials from the blood that the body can't use, much like you take out materials you can't use for garbage and recycling.

The job of some organs is to get rid of wastes.

The body's wastes move into the blood through the walls of the capillaries. As blood circulates throughout the body, it is filtered by not only the liver but also the kidneys. The kidneys are organs that remove wastes and extra water from the blood. If you followed a single drop of blood as it travels throughout the body, you would see that it moves through the liver and kidneys each time.

Your kidneys mix the wastes they filter from the blood with the water that is absorbed by the large intestine. This mixture is urine. The urine travels through long tubes to the bladder. Your bladder collects and stores the urine until it leaves your body.

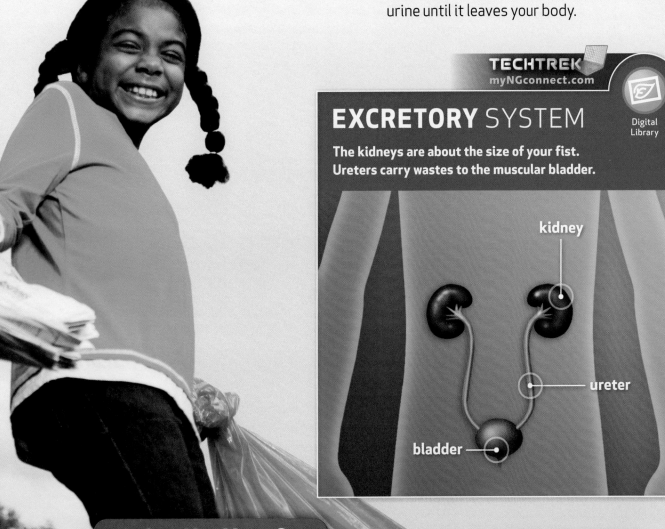

TECHTREK
myNGconnect.com
Digital Library

EXCRETORY SYSTEM
The kidneys are about the size of your fist. Ureters carry wastes to the muscular bladder.

- kidney
- ureter
- bladder

Before You Move On
1. What is the job of the digestive system?
2. What is the job of the excretory system?
3. **Draw Conclusions** Explain why your digestive and excretory systems could not do their jobs without the skeletal and muscular systems.

The Nervous System

Your nerves, spinal cord, brain, and sense organs make up your nervous system. These organs take in information and tell your body how to respond. Sometimes you are aware of this process. Sometimes you are not.

Nerves Nerves form a weblike network. They carry signals that allow you to sense the world around you and control your body. You do not always know about the messages that nerves carry.

Nerves help your body parts do involuntary jobs, such as digesting food and circulating blood.

Spinal Cord Your brain sends and receives information through your spinal cord. The spinal cord is a bundle of nerves. It directs signals from your brain to the other nerves that branch out into your body.

Your brain controls the actions of your body like this man controls the actions of the machine.

Brain The three main parts of your brain control different functions. The cerebrum is the largest part of the brain. It controls your voluntary muscle movements, your speech, and your senses. The cerebrum also enables you to think. It stores your memories and allows you to learn.

The cerebellum is located at the back of the head. This part controls balance and posture. It also fine-tunes movements so that the body can move gracefully.

The brain stem controls body functions that are automatic, such as breathing, swallowing, and your heartbeat.

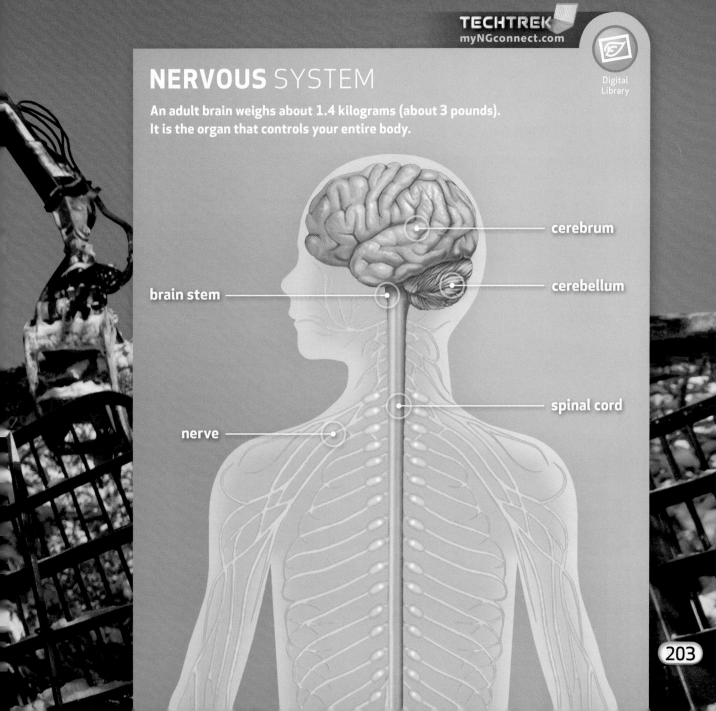

NERVOUS SYSTEM

An adult brain weighs about 1.4 kilograms (about 3 pounds). It is the organ that controls your entire body.

- cerebrum
- cerebellum
- brain stem
- spinal cord
- nerve

The Senses When you go to the movies, you can probably smell popcorn as soon as you walk into the theater! You see the action on the screen and hear the sound effects. As you eat your popcorn, you use your sense of taste. You use the sense of touch in your fingers to find and pick up the popcorn in the dark. You feel the cold of the lemonade you drink.

What is happening in these moviegoers' brains? They are using all of their sense organs at once. Sense organs are full of nerves that detect different kinds of information. They are the body's way of gathering information. You have sensory organs that deliver this information to your brain. Your brain uses information from your senses to determine action.

Scientists think the front part of your cerebrum controls your laughter!

SENSORY ORGANS

EYES Nerves in the eye detect light and color. The signals travel to your brain, which tells you that you see an object or a scene.

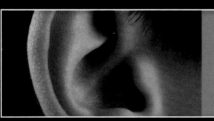

EARS Vibrations in the air move into the ear. Nerves detect the vibrations and send signals to your brain, which tells you that you hear sound.

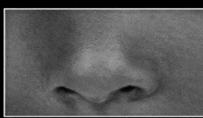

NOSE Very tiny particles in the air move into your nose. Nerves detect the particles and send signals to your brain, which tells you that you smell odors.

TONGUE Nerves in the tongue detect four tastes—sweet, sour, bitter, and salty. Nerves send signals about combinations of these to your brain, which tells you that you are tasting different flavors.

SKIN Nerves in the skin detect heat, cold, contact, pressure, and pain. The nerves send signals to your brain, which tells you what you feel.

Before You Move On

1. Name the human sensory organs.
2. Describe voluntary and involuntary ways that your brain is controlling your body right now.
3. **Analyze** How does the nervous system interact with the body's other systems?

Comparing Organ Systems

Like you, many other animals have spinal cords surrounded by protective bones. Fish, amphibians, reptiles, birds, and mammals are all vertebrates like humans. Although organs differ in structure among the groups, all have organ systems that carry out similar functions.

The nervous system of all vertebrates features an enlarged mass of nerve cells and sense organs at the front end of the spinal cord. This is the brain. In fish and amphibians, more of the brain is focused on the sense organs. In mammals, the cerebrum, where information is received and interpreted, is the largest part of the brain. That's why it is easier to train your dog to come when you call than your pet fish.

The dog has a larger cerebrum than other kinds of vertebrates.

Vertebrates eat a wide variety of foods. Flamingoes and some whales strain small crustaceans from the water through special mouthparts. Others eat seeds and other plant parts. Still others catch and kill large mammals. The digestive systems of individual vertebrates are specialized for the kind of food they eat.

Respiratory systems in vertebrates vary, but not as much as digestive systems. Generally, fish and young amphibians take in oxygen with gills. Reptiles, birds, and mammals have lungs. But oddities do exist. Many adult amphibians take in oxygen through their skin!

GETTING FOOD

GREAT WHITE SHARK Like other vertebrates that eat only meat, the digestive system of sharks does not have a very long intestine. They add strong digestive juices that act quickly to digest meat.

CARDINAL Birds have a two-sectioned stomach. One part is called the gizzard, which breaks down seeds. The gizzard is very muscular and helps grind up food. Because birds do not have teeth, they do not chew their food.

GETTING OXYGEN

PINEWOODS TREEFROG Most adult amphibians, such as pinewoods treefrogs, have lungs that are structured much like human lungs and do the same thing. All birds and reptiles have lungs, too.

HUMPHEAD CICHLID Fish, such as cichlids, have gills. Water moves in through the fish's mouth and out over the gills. Oxygen moves from the water into the fish's blood. Carbon dioxide moves from the blood into the water.

How does the food and oxygen get to the body in vertebrates? It travels in the circulatory system in much the same way as in humans. But the pathway is different in different vertebrates. In fish, the heart has only two chambers. It pushes blood in one big circle through the gills, to the body, and back to the heart for another push.

Vertebrates with lungs have a circulatory system divided into two circles. One circle leads from the heart to the lungs and back. The other leads from the heart to the body and back.

The excretory system of most vertebrates includes kidneys. But the kind of waste produced can be liquid, similar to yours, or solid crystals, as in birds and reptiles.

PUMPING BLOOD

SPOTTED SALAMANDER Like most reptiles and other amphibians, spotted salamanders have a three-chambered heart. Blood from the body and from the lungs enters the bottom chamber at the same time. Structures in that chamber direct most of the oxygen-rich blood to the body and the oxygen-poor blood to the lungs.

WOOD MOUSE Like other mammals, this wood mouse has a four-chambered heart. Blood follows the same pathway through the heart as it does in humans. Birds and crocodiles have four-chambered hearts, too.

Almost all vertebrates have bony skeletal systems. Some fishes, such as sharks, have skeletons made of cartilage, which is the same material that is in your nose and ears. Both kinds of skeletons provide support.

The muscular system carries out the same job in all vertebrates. Different kinds of muscles move bones, cause organs to move, and enable the heart to pump nonstop.

Fishes have muscles arranged in blocks that contract in waves so they can bend from side to side. Vertebrates with legs and wings have muscles that work in pairs like yours.

Birds have many bones with air spaces in them. The bones are lighter than they would be if they were solid. These light, strong bones and strong chest muscles enable birds to fly.

Before You Move On

1. What kinds of organ systems are found in all vertebrates?
2. How is the circulatory system of a fish different from a mammal?
3. **Apply** Crocodiles are reptiles that eat mostly meat. Tell how a crocodile's organ systems are like and different from your own.

NATIONAL GEOGRAPHIC

MAKING SENSE OF SENSES

Animals use their senses to keep track of their surroundings—to find food and to avoid threats. Many animals' senses are similar to your own. But their sensory organs can be very different from yours. The sharpness of animals' senses is different from yours, too.

Elephants have very sharp senses of hearing and smell. But their eyesight is not as good as a human's.

Hearing Elephants and bats both have much more sensitive ears than humans have, but in different ways. Elephants can detect sounds that are too low-pitched for humans to hear. They use these low-pitched sounds to communicate with one another over very long distances. Bats can detect sounds that are too high-pitched for humans to hear. They use sounds that people cannot hear to locate insects for food.

Touch Some animals detect objects using the sense of touch—but without actually touching the objects. A cat's whiskers detect changes in air movements near an object. A seal's whiskers work the same way, but they detect movements in water. Insects, crabs, and lobsters have antennae that feel both objects and motion.

The big brown bat tracks insects using sound instead of sight.

Whiskers help cats feel their way in very low light.

Taste Catfish have antennaelike structures called barbels. Barbels are covered with taste sensors. Catfish do not see well, so they taste their way along the bottoms of rivers and lakes. Some types of blind fish have taste sensors covering most of their bodies. The bodies of earthworms are covered with taste sensors, too. You only taste with your tongue, but an octopus tastes with its tentacles. A housefly tastes with its feet!

Smell When a snake flicks its tongue, it's not to grab food. It's actually smelling the environment. The tongue brings tiny air particles into an organ on the roof of its mouth that interprets the odor. Many animals have a much stronger sense of smell than humans. Some breeds of dogs can detect where a human simply walked a few days ago. You might be able to smell your lunch from across the room, but a grizzly bear can smell food buried deep in the ground. Some bears can detect scents in the air coming from over 24 kilometers (about 15 miles) away!

This green ratsnake is sensing the environment.

Catfish barbels have many more taste sensors than your tongue.

Sight You have two eyes in the front of your head, but you can turn your head to get a better look at something. A chameleon just points its two eyes in different directions! Eagles, hawks, and owls have sharp eyesight and can spot small prey from high in the air. Spiders have as many as eight eyes bunched around the top of the head looking in all directions. Animals that spend their entire lives in dark caves or in deep water might not have eyes of any sort.

This tawny owl hunts well at night because its very large eyes work together to see objects in 3-D like you.

This Usambara giant three horn chameleon can look in more than one direction at a time.

Conclusion

Teams of organs work together in humans to carry out certain functions. The skeletal and muscular systems enable support and movement. The circulatory, respiratory, and digestive systems work together to supply nutrients and oxygen to the body and carry away carbon dioxide. They also enable the excretory system to remove wastes. All these efforts are coordinated by the nervous system. Human organ systems can be compared with those of other vertebrates, where the same job might be carried out with similar or very different organs.

Big Idea Organ systems work together to perform specific functions in humans and other animals.

Vocabulary Review

Match each of the following terms with the correct definition.

A. nervous system
B. digestive system
C. excretory system
D. respiratory system
E. circulatory system
F. organ system

1. A group of organs that works together to do a specific job in the body
2. A group of organs that breaks food down into nutrients the body can use
3. A group of organs that carries blood and oxygen to the body and wastes away
4. A group of organs that takes in information from the surroundings and tells the body how to respond
5. A group of organs that takes oxygen into the body and removes wastes
6. A group of organs that removes wastes from the blood that the body cannot use

Big Idea Review

1. **Define** How does an organ differ from an organ system?

2. **List** Which organ systems are involved in removing wastes from the body?

3. **Cause and Effect** How do bones and muscles work together when you raise your arm?

4. **Compare and Contrast** List features of organ systems in vertebrates. By each feature, write the name of the kind of vertebrate that uses that feature.

5. **Draw Conclusions** Does any organ system act alone as it does its specific job? Explain.

6. **Evaluate** Do you think the heart or brain is the control center of the body? Support your answer.

Write About Organ Systems

Explain What is happening in this photo? What organ systems are working together to make this action possible?

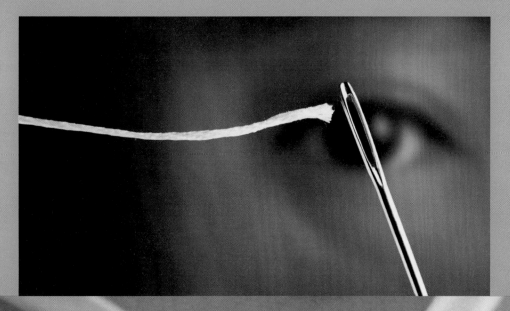

NATIONAL GEOGRAPHIC

CHAPTER 5 — LIFE SCIENCE EXPERT: SURGEON

Dr. Arviso Alvord works closely with people in all parts of her job.

Dr. Lori Arviso Alvord is a Navajo surgeon. She received her training at Stanford Medical School. She combines traditional Navajo healing with her training to treat patients.

What do you do as a surgeon?

I work on the inside of people's bodies and fix problems they are having. For example, I might remove an appendix or a gallbladder if it is causing problems. I am trained to do many different operations on the organs of the body. Sometimes, when I operate, it saves a person's life.

What do you remember liking about science when you were in school?

Science is all about discovery and understanding the world. I loved the study of animals and how they live in their environment. I also enjoy any kind of chemistry experiment that makes something cool happen!

What's a typical day like for you?

I spend most days either in the clinic or in the operating room. On clinic days, we start at around 8 a.m., end around 5 p.m., and see patients all day. We figure out what sort of surgical problem someone has, and then often schedule surgery. Sometimes we are called to the emergency room to see a patient who needs surgery right away. On an operating room day, we spend all day operating. We might do three to five operations. We are part of a team that works closely together all day long.

What has been the coolest part of your job?

The first really cool part is doing an operation. You get to work with your hands. You do a lot of dissection and clamping and tying of blood vessels. Sometimes you operate using instruments through small openings in the skin, called laparoscopic surgery. This is both fun and interesting.

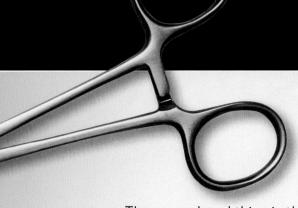

The second cool thing is that you do something that makes a huge difference in someone's life. It is an honor to be able to help people so very much.

What has been your greatest accomplishment so far?

My greatest accomplishment so far has been my ability, as a surgeon, to make a difference in people's lives, to relieve their pain, and to help them to get well. Sometimes I can save a life. But I've also had success as an author and a public speaker. I have served as a role model for many Native American children. I hope I have shown them that they can do whatever they set their minds to do. I also hope that my work to create "healing environments" for patients, by using Native American philosophies of healing, has helped to move medicine in a direction that helps all patients.

What advice would you give young people who want to become surgeons?

Learn all you can about medicine and surgery and be sure that's what you want to do. Then do things that develop your mind, such as music, art, reading and literature, athletics and dance. Also, since medicine is all about helping people, develop the right attitude toward people. You can start by getting involved in community service projects.

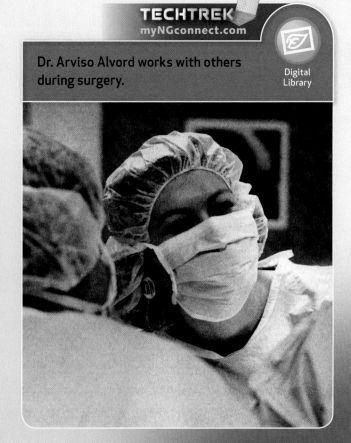

Dr. Arviso Alvord works with others during surgery.

NATIONAL GEOGRAPHIC

BECOME AN EXPERT

Ironman Triathlon Race: Organ Systems Working Together

A triathlon is a race of three parts: swimming, bicycling, and running. Triathlons are different lengths, but the hardest is the Ironman Triathlon. Racers have to swim 3.8 km (2.4 mi), bike for 180 km (112 mi), and then run for 42.2 km (26 mi).

Racers have to train incredibly hard to prepare their bodies for such a long race. By the last part of the race, their bodies are already exhausted even if all factors, such as the weather, are perfect.

Hard work takes place in the racer's organs. The heart pumps blood; the lungs take in oxygen; muscles pull on bones to move the body forward. All the **organ systems** have to work together for the racer to finish this incredible task.

Racers start the first part of the race—swimming.

organ system

An **organ system** is a group of organs that works together to do a specific job in the body.

Energy for the Body

Triathletes need a great deal of energy for over twelve hours of racing. The energy comes from food.

The **digestive system** turns food into nutrients the body can use. Digestion begins in the mouth, where teeth chew food and mix it with saliva. Food moves to the stomach and digestive juices are added. Then the food becomes a thick liquid and moves into the small intestine, where digestion is finished. Juices from the pancreas and the liver help break down food into nutrients. The small intestine absorbs the nutrients into the blood. Any remaining food passes to the large intestine, where water is absorbed.

digestive system
The **digestive system** is a group of organs that breaks down food into nutrients the body can use.

BECOME AN EXPERT

Getting Food and Oxygen

Training strengthens organ systems for high performance. The **circulatory system** must be strong enough to pump blood—carrying nutrients from the digestive system and removing waste products at the same time—throughout the grueling race.

Blood travels in a circlelike path that has no beginning and no end. If you start on the left side of the heart, blood rich in oxygen moves through arteries to the body. Blood reaches the capillaries, where oxygen moves out of the capillaries into the body. Carbon dioxide then moves back into the blood. Food and waste products are also exchanged. Then, through the veins, the blood moves back to the right side of the heart.

Most racers take a little over an hour to complete the swimming part. Their blood probably cycled at least 60 times.

circulatory system

The **circulatory system** is a group of organs that carries blood and oxygen to the body and wastes away.

Then the circulatory system interacts with the **respiratory system**. From the right side of the heart, blood travels to the lungs. Here the carbon dioxide is exchanged for oxygen, and the blood moves back to the left side of the heart. The cycle repeats hundreds of times during the race.

The racer has to inhale and exhale just right so only air enters the lungs. Breathing starts at the nose and mouth. The racer inhales air. A muscle called the diaphragm, at the base of the rib cage, flattens and lets the lungs expand when the athlete inhales. The lungs fill with air. When the racer exhales, the diaphragm relaxes and the lungs empty.

This racer exhales underwater.

respiratory system
The **respiratory system** is a group of organs that takes oxygen into the body and removes wastes.

BECOME AN EXPERT

Support and Movement

Just like the bicycles they are riding, racers need strong frames. The skeletal and muscular systems work together to support racers over the long distance they must ride.

The racer's skeleton flexes and bends at its joints to pedal the bicycle. The many small bones of the backbone enable the bicyclist to bend over smoothly to reach the handlebars. The elbows bend slightly. And the joints between the many bones in the wrists and hands are always flexing to ensure the right grip.

The bicycling part takes about six hours for most racers. Think about how many times a thigh muscle would contract during that time!

The skeletal muscles, or those muscles attached to bones, are working continuously during all parts of the race. Some muscles are used more in one part of the race than in others. Bicycling uses the muscles of the legs and lower body. Even the back muscles are contracting to keep the racer hunched down over the bicycle. Staying hunched over as much as possible means the racer is more streamlined and can ride faster.

Other kinds of muscles are at work, too. The cardiac muscle of the heart is contracting at a fast pace. The smooth muscles of the arteries are helping get the blood to the body. The diaphragm moves up and down in a regular pattern to get enough air into the body.

HOW BONES AND MUSCLES WORK TOGETHER

Muscles contract, or pull. Muscles often work in pairs to move bones. Notice which muscle contracts and which relaxes to pedal a bicycle.

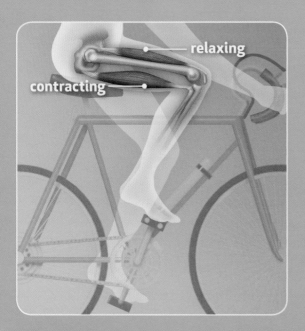
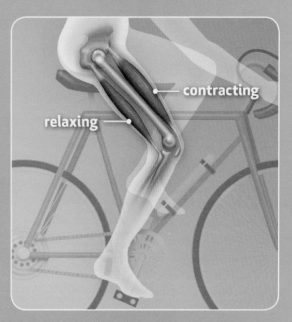

BECOME AN EXPERT

Control and Senses The work of the racer's body must be coordinated and controlled. The **nervous system** commands the functions of the entire body. The brain, spinal cord, and nerves move messages extremely quickly to and from all body parts.

The nerves form a web throughout the body. They carry messages to the spinal cord. From there the messages speed to the brain. The brain processes the information and responds. It sends orders back through the spinal cord to the nerves.

Running might seem somewhat automatic. But after the swimming and bicycling, it is not easy to continue on foot. The racer must be able to focus the brain on not tripping, especially if there's a small dip in the pavement. And that's in addition to what the brain is controlling already!

nervous system

The **nervous system** a group of organs that takes in information from the surroundings and tells the body how to respond.

The body and brain respond to the environment. The racer gathers information about the temperature of water and air, the motion of waves, noise, and the location of holes along the side of the road. Sensory organs—the eyes, ears, nose, tongue, and skin—collect this information for the brain.

The skin does several other jobs during the race. One of the most important is keeping the racer's body at the right temperature. It also helps protect the body if the racer falls.

Most racers take about four and one-half hours to complete the running part. It can take longer when the weather is hot.

BECOME AN EXPERT

Removing Waste A racer's body produces waste as it works. The **excretory system** ensures all of the organs are constantly cleaned. The liver and kidneys filter waste from blood. The two kidneys filter all the body's blood about two times each hour. They return clean blood to the blood vessels and mix the waste with water. The liquid waste, urine, goes to the bladder.

An exhausted racer barely crossed the finish line.

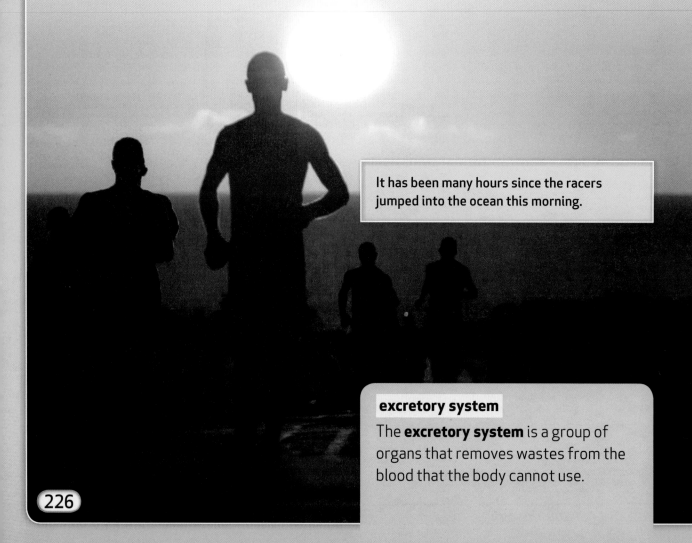

It has been many hours since the racers jumped into the ocean this morning.

excretory system
The **excretory system** is a group of organs that removes wastes from the blood that the body cannot use.

Success! What a fatiguing race! The winner might finish in less than nine hours. Most racers take longer. But the organs in every racer have to be working their best. The triathlon pushes every organ to its limit. Strength training for organs and training them to work over long periods of time are very important for success. And each racer has to have a winning attitude!

Imagine being able to smile after winning this exhausting race!

BECOME AN EXPERT

CHAPTER 5: SHARE AND COMPARE

Turn and Talk How do organ systems work together in living things? Form a complete answer to this question together with a partner.

Read Select two pages in this section that are the most interesting to you. Practice reading the pages so that you can read them smoothly. Then read them aloud to a partner or small group. Talk about why the pages are interesting.

Write Write a conclusion that tells the important ideas you have learned about how organ systems work together in living things. State what you think is the Big Idea of this section. Share what you wrote with a classmate. Compare your conclusions. Did your classmate recall that muscles work in pairs to move bones?

Draw Imagine what it is like racing in a triathlon. Draw a picture of an organ system that helps a racer during a triathlon. Combine your drawing with some of your classmates' in a triathlon racer's body outline. Present your group's racer by telling how the organ systems worked and if your racer won!

NATIONAL GEOGRAPHIC

Glossary

B

behavior (bē-HĀV-yur)
Behavior is any way that an animal interacts with its environment. (p. 144)

C

cell (SEL)
A cell is the smallest unit of a living thing. (p. 11)

Plants are made up of cells.

chlorophyll (KLOR-uh-fil)
Chlorophyll is the green material in plants that absorbs sunlight, which the plant uses to make food. (p. 108)

circulatory system (SIR-kyū-luh-tor-ē SIS-tum)
The circulatory system is a group of organs that carries blood and oxygen to the body and wastes away. (p. 192)

communication (kuh-MYŪ-ni-kĀ-shun)
Communication is any behavior that lets animals share information. (p. 150)

community (kuh-MYŪ-nuh-tē)
A community is all the different organisms that live and interact in an area. (p. 61)

D

digestive system (dī-JES-tiv SIS-tum)
The digestive system is a group of organs that breaks down food into nutrients the body can use. (p. 198)

Bird songs are one kind of communication in the animal world.

GLOSSARY

E

ecosystem (Ē-kō-sis-tum)
An ecosystem is all the living things and the nonliving things in an area and their interactions. (p. 58)

excretory system (EKS-kre-tor-ē SI-stem)
The excretory system is a group of organs that removes wastes from the blood that the body cannot use. (p. 200)

F

food chain (FŪD CHĀN)
A food chain is a process by which energy passes from one living thing to another. (p. 116)

food web (FŪD WEB)
A food web is a process that combines many food chains to show how energy moves through an ecosystem. (p. 118)

H

habit (HA-bit)
A habit is a behavior that is learned through practice. (p. 147)

I

instinct (IN-stinkt)
An instinct is an inherited behavior that an animal can do without ever learning how to do it. (p. 144)

Wasps build their nest by instinct.

invertebrate (in-VUR-tuh-brit)
An invertebrate is an animal with no backbone. (p. 20)

This butterfly is an invertebrate because it does not have a backbone.

GLOSSARY

K

kingdom (KING-dum)
A kingdom is one of the six main groups into which living things are sorted. (p. 12)

Mushrooms are classified in the fungi kingdom.

L

learning (LUR-ning)
Leaning is a change in behavior that comes about through experience. (p. 145)

N

nervous system (NER-vus SIS-tum)
The nervous system is a group of organs that takes in information from the surroundings and tells the body how to respond. (p. 202)

niche (nēsh)
A niche is the way an organism interacts with living and nonliving parts of its ecosystem. (p. 62)

O

organ system (OR-gun SIS-tum)
An organ system is a group of organs that works together to do a specific job in the body. (p. 187)

P

photosynthesis (FŌ-tō-SIN-thuh-sis)
Photosynthesis is the process by which green plants make food by using energy from sunlight. (p. 108)

population (pop-yū-LĀ-shun)
A population is all the members of a species living in a certain area. (p. 60)

R

respiratory system (RES-pur-uh-tor-ē SIS-tum)
The respiratory system is a group of organs that takes oxygen into the body and removes wastes. (p. 196)

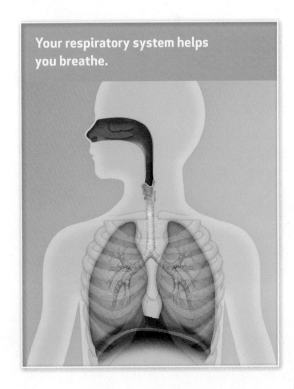

Your respiratory system helps you breathe.

GLOSSARY

S

species (SPĒ-sēz)
A species is a group of similar living things that can produce offspring who can also produce offspring. (p. 14)

succession (suhk-SESH-un)
Succession is the gradual replacement of one community by another over time. (p. 71)

symbiosis (sim-bē-Ō-sis)
Symbiosis is the close association of members of different species that live together. (p. 66)

V

vascular plant (VAS-kyū-lar PLANT)
A vascular plant is one that contains bundles of tubelike cells that transport water and food throughout the plant. (p. 17)

In these vascular plants, water is transported from the roots to the leaves and flowers.

vertebrate (VUR-tuh-brit)
A vertebrate is an animal with a backbone. (p. 28)

This species of camel has two humps.

Index

A

Acid rain, 81

Air pollution, 80–83

All Hands on Deck! Saving Right Whales, 152–155

Analyze, 87, 123, 205

Animal adaptations
  inherited traits, 138, 144, 146, 148–149, 151
  for survival and reproduction, 44–51, 138, 144, 146, 148–149, 151, 153, 158–159

Animals,
  cells, 11
  characteristics of, 138–139, *chart, 139,* 144–145, 159–160, 162–163, 172–179
  classification of, 12–13, 14–15, 20–37
  as consumers, 107, 110–113, 117, 126–127
  extinction of, 164–165
  food, plant origins of, 107, 109
  frightful, 44–51
  invertebrates, 20–27
  as predators, 62–63
  protection of, right whales, 152–155
  ranges of, *maps, 46–51*
  senses, 210–213
  shelters, 148–149, *chart, 149*
  vertebrates, 28–37, 206–209

Ants, 172–179

Apply, 19, 41, 113, 159, 187, 209

Aquatic ecologist, 170–171

Arctic ecosystems, 126–131
  food chain, 128
  food web, *diagram, 130–131*

Arviso Alvord, Lori, 216–217

B

Bacteria and fungi, 114–115

Become an Exert
  Amazing World of Ants, The, 172–180
  Energy in Ecosystems: The Arctic, 126–132
  Frightful Animals: Just Trying to Survive, 44–52
  Interactions in a Coral Reef, 90–100
  Ironman Triathlon Race: Organ Systems Working Together, 218–228

Before You Move On, 15, 19, 27, 37, 61, 65, 69, 75, 83, 109, 113, 115, 119, 143, 151, 159, 167, 187, 191, 197, 201, 205, 209

Behavior
  big idea review, 169
  communication, 150–151
  conclusion, 168
  cooperative, 148, 172–179
  definition of, 136, 144, 172
  as environmental interaction, 145, 147
  instinctive, 44–51, 144, 146, 148–149
  learned, 145, 147
  social, of ants, 172–179
  vocabulary review, 168

Big Idea Questions
  How Do Body Systems Work Together? 181–228
  How Does Energy Move in an Ecosystem? 101–132

INDEX

How Do Living Things Survive and Change? 133–180

How Do Scientists Classify Living Things? 5–52

What Are the Interactions in Ecosystems? 53–100

Big Idea Review, 41, 87, 123, 169, 215

Biographical information on scientists. *See* Life Science Expert; Meet a Scientist

Body systems, 182–185. *See also* Organ system(s)

Bones. *See* Skeletal system

Butterflies, monarch, 158–159, *diagram of life cycle,* 159

C

Canopy biologist, 42–43

Careers, science. *See* Life Science Expert; Meet a Scientist

Carnivores, 111, 117, 126–129

Cause and effect, 169, 215

Cell
definition of, 8, 11, 51
vocabulary review, 40

Cells
animal and plant, compared, 11
basic structures and functions of, 11, *diagram,* 11, 12–13, 16–17, 20–23
organisms' smallest units, 11

Characteristics, acquired
of animals, 145, 147
environmental influences on, 142–143, 145
of plants, 142

Characteristics, inherited
of animals, 138–139, 144, 146, 148–155, 159, 160, 162–163, 172–179
behavioral, 144, 146–149, 172–179

physical, 138–141, 152–153, 160–163
of plants, 140–141, 156, 161

Charts
animal shelters, 149
classification of a bactrian camel, 15
comparing animal feet, 139
dichotomous key, 27
flower color and pollinators, 141
getting food, 207
getting oxygen, 207
kingdoms, characteristics of, 12–13
pumping blood, 208
sensory organs, 205

Chlorophyll, 104, 108, 122, 128

Circulatory system
compared among vertebrates, 208
conclusion, 214
definition of, 184, 192, 220
general purpose of, 192–195, *diagram,* 193
human heart, 194–195, *diagram,* 195
and the Ironman Triathlon Race, 220
vocabulary review, 214

Classification
dichotomous key, using a, 27
groups, named, 14
of invertebrates, 20–27
kingdoms, 12–13
levels within kingdoms, 14–15, *chart,* 15
of living things, 10–15
of plants, 16–19
role of cells in, 11, 12–13, 16–17, 20–23
by shared characteristics, 10–11
of vertebrates, 28–37
write about, 41

Classify, 37, 41, 123

INDEX

Communication
    in ant colonies, 178–179
    big idea review, 169
    definition of, 137, 150, 178
    for protection, 150
    for reproduction, 151
    vocabulary review, 168

Communities
    big idea review, 87
    changes due to succession, 70–73
    conclusion, 86
    coral reef, 90

Community
    definition of, 56, 61, 90
    vocabulary review, 86

Compare and contrast, 215

Competition, 62–65

Conclusion, 40, 86, 122, 168, 214

Conservationist, 88–89

Consumers
    in the Arctic food web, 126–129
    big idea review, 123
    carnivores, 111, 117, 118–119
    classified by, 110–113
    conclusion, 122
    deep-sea creatures as, 120–121
    definition of, 107
    herbivores, 110, 116, 118–119
    omnivores, 110, 112–113, 118–119
    organisms supplying energy to, 107–109

Contrast, 41, 87

Cooperative behaviors, 148, 172–179

Coral reef
    availability of food and resources in, 90–96
    changing physical characteristics of, 90–99
    conserving and restoring a, 98–99
    organisms' interactions in, 92–95
    patterns of behavior in, 90–95

D

Decomposers
    bacteria and fungi, 114–115
    recycling nutrients into the soil, 115

Deep-Sea Vents: Living on the Edge, 120–121

Deer ears, modeling, 139

Define, 123, 215

Desert food chain, *diagram,* 116–117

Desert food web, *diagram,* 118–119

Describe, 169

Diagrams
    Arctic food web, 130–131
    bones' spongy part, 188
    circulatory system, 193
    desert food chain, 116–117
    desert food web, 118–119
    digestive system, 199
    excretory system, 201
    heart's chambers, 195
    how bones and muscles work together, 223
    hyacinth, yearly cycle of, 157
    monarch, life cycle of, 159
    muscular system, 191
    nervous system, 203
    photosynthesis, 108
    plant cell, 11
    respiratory system, 196
    skeletal system, 188

INDEX

Dichotomous key, 27

Digestive system
    compared among vertebrates, 207
    conclusion, 214
    definition of, 185, 198, 219
    general purpose of, 198–199, *diagram,* 199
    and the Ironman Triathlon Race, 219
    vocabulary review, 214

Digital Library, 5, 7, 21, 29, 31, 43, 45, 53, 55, 63, 65, 66, 88, 89, 91, 97, 103, 111, 125, 127, 129, 133, 135, 141, 149, 171, 173, 174, 183, 188, 203, 217, 219, 222

Discovering a Species in Madagascar, 38–39

Draw conclusions, 15, 65, 109, 119, 123, 201, 215

E

Ecologist, 121, 124–125

Ecosystem
    Arctic, 126–131
    balanced, 63, 76
    big idea review, 87
    competition in, 62–65
    composed of living and nonliving things, 58–59
    conclusion, 86
    coral reefs, example of, 90–99
    definition of, 56, 59, 90
    energy in (*See* Energy in an ecosystem)
    human's changes to, 75–85, 88–89, 96–99
    invasive species' changes to, 74–75
    meeting organisms' basic needs, 58–59
    meeting people's basic needs, 76–79
    organism's changes to, 62–65, 70–75, 90–99
    pollution's effect on, 78–81, 97
    predation and competition's, effect on, 62–63
    succession's effect on, 70–73
    vocabulary review, 86

Energy in an ecosystem
    movement from sunlight to decomposers, 114–115
    movement through food chains, 116–117, 126–127
    movement through food webs, 118–119, 120–121, 126–127
    for survival of all living things, 106–107
    write about, 123

Enrichment activities, 5, 7, 14, 19, 53, 55, 103, 118, 133, 135, 159, 183, 199

Ethnobotanist, 4

Evaluate, 27, 61, 215

Evolution
    and fossils' record of past life, 166
    of living fossils, 167
    as response to environmental change, 156–163

Excretory system
    big idea review, 215
    compared among vertebrates, 208
    conclusion, 214
    definition of, 185, 200, 226
    general purpose of, 200–201, *diagram,* 201
    and the Ironman Triathlon Race, 226
    vocabulary review, 214

Explain, 41, 87, 123, 169, 215

Extinction
    due to catastrophic events, 166
    fossil evidence of, 166
    of plants and animals, 164–165

INDEX

protecting animals from, 152–155

as a result of environmental change, 166

F

Fadiman, Maria, 4

Food chain

in the Arctic, 126–131

conclusion, 122

definition of, 105, 116, 127

of desert organisms, 116–117

phytoplankton in, 127–128

sun's energy moving through, 116–117, 127–128

vocabulary review, 122

Food web

in the Arctic, 126–131, *diagram, 130–131*

conclusion, 122

in the deep ocean, 120–121

definition of, 105, 118, 127

of a desert ecosystem, 118–119

and laboratory studies, 124–125

vocabulary review, 122

Fossils

as evidence of past life, 166

living, 167

vocabulary review, 122

G

Generalize, 41, 83, 115, 191, 197

Glossary, EM1–EM3

H

Habit, 137, 147, 168

Herbivores, 110, 116

Hogan, Zeb, 170–171

Hyacinth, yearly cycle of, *diagram,* 157

I

Identify, 123

Infer, 69, 151, 169

Instinct

conclusion, 168

definition of, 136, 144, 173

as inherited characteristic, 144, 146, 148, 149, 173

as protection, 146–147, 150, 175

vocabulary review, 168

Invertebrate

definition of, 9, 20, 46

vocabulary review, 40

Invertebrates

arthropods, 25

big idea review, 41

EM9

INDEX

classifying, 20–27
conclusion, 40
insects, 26–27
jellies, anemones and corals, 21
sea stars, 24
snails and octopuses, 23
sponges, 20
worms, 22

Ironman Triathlon Race, 218–227

K

Kingdom
characteristics of, 12–13
classification levels within, 14–15
conclusion, 40
definition of, 8, 12, 44
vocabulary review, 40

L

Learning
conclusion, 168
definition of, 136, 145, 179
flexibility of, 145
from parents, 145
vocabulary review, 168

Life cycles
as adaptations to environment, 156–159
of ants, 174, 176–177
of monarch butterflies, 158–159
plants' survival adaptations, 156–157

Life Science Expert
aquatic ecologist, 170–171
Arviso Alvord, Lori, 216–217
canopy biologist, 42–43
conservationist, 88–89
ecologist, 124–125

Hogan, Zeb, 170–171
López-Mendilaharsu, Milagros, 88–89
Lowman, Meg, 42–43
Martinez, Neo, 124–125
surgeon, 216–217

List, 41, 87, 169, 215

López-Mendilaharsu, Milagros, 88–89

Lowman, Meg, 42–43

M

Make judgments, 87, 169

Making Sense of Senses, 210–213

Maps
blood python, range of, 50
blue-ringed octopus, range of, 51
European eagle owl, range of, 47
Gila monster, range of, 45
great white shark, range of, 49
North America's East Coast, 153
spotted hyena, range of, 48
tarantula spider, range of, 46

Martinez, Neo, 124–125

Meet a Scientist
ethnobotanist, 4
Fadiman, Maria, 4

Migration, 149, 153–155, 158

Monarch butterflies, life cycle of, 158–159

Muscular system
big idea review, 215
compared among vertebrates, 209
conclusion, 214
definition of, 190
general purpose of, 190–191, *diagram,* 191
and the Ironman Triathlon Race, 222–223

INDEX

muscles and bones working together, 223

muscles, voluntary and involuntary, 190–192, 196

My Science Notebook

draw, 52, 100, 132, 180, 228

write, 52, 100, 132, 180, 228

write about animal behavior, 169

classification, 41

energy in an ecosystem, 123

interactions, 87

organ systems, 215

N

Nervous system

compared among vertebrates, 206

conclusion, 214

definition of, 185, 202, 224

general purpose of, 202–203, *diagram*, 203

and the Ironman Triathlon Race, 224–225

vocabulary review, 214

Niche

conclusion, 86

definition of, 57, 62, 95

and role of predators, 62–63

of species on a coral reef, 94–95

vocabulary review, 86

Nonvascular plants, 17

North America's East Coast, *map*, 153

O

Omnivores, 112–113

Organs

bladder, 210, 226

brain, 202–203, 224–225

heart, 187, 191, 192, 194–195, 220

intestines, 198–199, 219

kidneys, 210, 226

liver, 199, 219, 226

lungs, 187, 196–197, 220

nerves, 202–205, 224

pancreas, 199, 219

sensory organs, 204–205, 225

skin, 186, 225

spinal cord, 202, 224

stomach, 198, 219

Organ system(s). *See also* Organs

big idea review, 215

circulatory, 192–195, *diagrams*, 193, 195, 220

compared among vertebrates, 206–209

conclusion, 214

definition of, 184, 187, 214, 218

digestive, 198–199, *diagram*, 199, 219

excretory, 200–201, *diagram*, 201, 226

general purpose of, 186–187

muscular, 190–191, *diagram*, 191, 222–223

nervous, 202–203, *diagram*, 203, 224

respiratory, 196–197, *diagram*, 196, 221

sensory, 204–205, *chart*, 205, 210–213, 225

skeletal, 188–189, diagram, *188*, 222–223

vocabulary review, 214

working together: Ironman Triathlon Race, 218–227

write about, 215

P

Photosynthesis

in the Arctic, 127–128

conclusion, 122

definition of, 104, 108, 127

EM11

INDEX

transferring sunlight to chemical energy, 108–109, *diagram*, 108

vocabulary review, 122

Plant adaptations

flower color and pollinators, 141

for survival in changing environments, 156–157

Plants

cells, 11

characteristics of, 140–142, 156, 161

classification of, 16–19

conifers, 19

consumers' use of, 106–107, 110–111, 116–118

extinction of, 165

ferns, 19

flowering, 18–19

mosses, 17

as producers, 106, 108–109, 117

use of sunlight by, 106–109

vascular and nonvascular, 16–17

Population

of beaver and ecosystem change, 70–71

big idea review, 87

and communities, 60–61

of coral, and reef development, 91, 94–95

definition of, 56, 60, 91

vocabulary review, 86

Predation

and competition, 62–65

and population control, 63, 74, 75

predators' niche, 62–63

Predict, 75, 87, 143, 167

Producers

in the Arctic food web, 126–127

big idea review, 123

conclusion, 122

deep-sea bacteria as, 120

definition of, 106

as first food chain organism, 117

plants and algae as, 106

sunlight supplying energy to, 106–109

R

Recall, 41, 87

Relate, 169

Reproduction of vascular plants, 18–19

Respiratory system

compared among vertebrates, 207

conclusion, 214

definition of, 184, 196, 221

general purpose of, 196–197, *diagram*, 196

and the Ironman Triathlon Race, 221

vocabulary review, 214

Right whales, 152–155

Rooftop Gardens, 84–85

INDEX

S

Science in a Snap!
Using a Dichotomous Key, 27
Feel the Beat, 194
Model Deer Ears, 139
Personal Impact, 83
What Have You Eaten? 113

Science Vocabulary, 8–9, 56–57, 104–105, 136–137, 184–185
behavior, 136, 144, 168, 172
cell, 8, 11, 40, 51
chlorophyll, 104, 108, 122, 128
circulatory system, 184, 192, 214, 220
communication, 137, 150, 168, 178
community, 56, 61, 86, 90
digestive system, 185, 198, 214, 219
ecosystem, 56, 59, 86, 90
excretory system, 185, 200, 214, 226
food chain, 105, 122, 127
food web, 105, 122, 127
habit, 137, 147, 168
instinct, 136, 144, 168, 173
invertebrate, 9, 20, 40, 46
kingdom, 8, 12, 40, 44
learning, 136, 145, 168
nervous system, 185, 202, 214, 224
niche, 57, 62, 86, 95
organ system, 184, 187, 214, 218
photosynthesis, 104, 108, 122, 127
population, 56, 60, 86, 91
respiratory system, 184, 196, 214, 221
species, 8, 14, 40, 44
succession, 57, 71, 86, 94
symbiosis, 57, 66, 86, 92
vascular plant, 9, 17, 40, 50
vertebrate, 9, 28, 40, 48

Scientists. *See* Life Science Expert; Meet a Scientist

Senses among animals
general purpose of, 210
hearing, 211
sight, 213
taste, 212
touch, 211

Sensory organs
definition of, 204
general purpose of, 204–205
and the Ironman Triathlon Race, 224–225
senses, among animals, 210–213
senses, human, 204–205
vertebrate comparisons, 210–213

Share and Compare, 52, 100, 132, 180, 228

Skeletal system
big idea review, 215
bones and muscles working together, 223
compared among vertebrates, 209
conclusion, 214
definition of, 188
general purpose of, 188–189, *diagram, 188*
and the Ironman Triathlon Race, 222–223

Species
conclusion, 40
definition of, 8, 14, 44
discovery of a, 38–39
as a level in classification, 14–15
vocabulary review, 40

Student eEdition, 5, 7, 43, 45, 53, 55, 89, 91, 103, 125, 127, 133, 135, 171, 173, 183, 217, 219

INDEX

Succession
    beaver dams' effect on, 70–71
    big idea review, 87
    conclusion, 86
    on coral reefs, 94–95
    definition of, 57, 71, 94
    forest fires' effect on, 72–73
    vocabulary review, 86

Surgeon, 216–217

Symbiosis
    beneficial to both organisms, 66–67, 92–93
    big idea review, 87
    conclusion, 86
    on a coral reef, 92–93
    definition of, 57, 66, 92
    one-way relationships, 68
    parasites, 69
    vocabulary review, 86

INDEX

T

Tech Trek
digital library, 5, 7, 21, 29, 31, 43, 45, 53, 55, 63, 65, 66, 88, 89, 91, 97, 101, 103, 111, 125, 127, 129, 133, 135, 141, 149, 171, 173, 174, 183, 188, 203, 217, 219, 222

enrichment activities, 5, 7, 14, 53, 55, 101, 103, 118, 133, 135, 159, 183, 199

student eEdition, 5, 7, 43, 45, 53, 55, 91, 101, 103, 125, 127, 133, 135, 171, 173, 183, 217, 219

vocabulary games, 5, 7, 9, 53, 55, 57, 101, 103, 105, 133, 135, 137, 183, 185

Thinking Skills
analyze, 87, 123, 205
apply, 19, 41, 113, 159, 187, 209
cause and effect, 169, 215
classify, 37, 41, 123
compare and contrast, 215
contrast, 41, 87
define, 123, 215
describe, 169
draw conclusions, 15, 65, 109, 119, 123, 201, 215
evaluate, 27, 61, 215
explain, 41, 87, 123, 169, 215
generalize, 41, 83, 115, 191, 197
identify, 123
infer, 69, 151, 169
list, 41, 87, 169, 215
make judgments, 87, 169
predict, 75, 87, 143, 167
recall, 41, 87
relate, 169

V

Vascular plant
definition of, 9, 17, 50
vocabulary review, 40

Vascular plants
definition of, 9, 17, 50
groups of, 18–19
reproduction of, 18–19
vocabulary review, 40

Vertebrate
definition of, 9, 28, 48
vocabulary review, 40

Vertebrates
amphibians, 30–31
big idea review, 41
birds, 34–35
classifying, 28–37
fishes, 29
mammals, 36–37
organ systems compared, 206–213
reptiles, 32–33
structures and functions compared, 28–37

Vocabulary games, 5, 7, 9, 53, 55, 57, 105, 133, 135, 137, 183, 185

Vocabulary review, 40, 86, 122, 168, 214

W

Water pollution, 78–79, 97

What Is Life Science? 2–3

Write about
animal behavior, 169
classification, 41
energy in an ecosystem, 123
interactions, 87
organ systems, 215

EM15

Credits

Front Matter
About the Cover (b inset) D. Parer & E. Parer-Cook/Auscape/Minden Pictures. (bg) Stephen Alvarez/National Geographic Image Collection. (t inset) Stephen Alvarez/National Geographic Image Collection. **ii-iii** Pixtal Images/Photolibrary. **iv-v** John Shaw/Photoshot. **ix** (inset) Stuart Westmorland/Image Bank/Getty Images. **vii** (t) Top-Pics TBK/Alamy Images. **vii-ix** (bg) Darlyne A. Murawski/National Geographic Image Collection. **vi-vii** (bg) Jonathan Blair/National Geographic Image Collection. **x-1** Ludmila Yilmaz/Shutterstock. **2** (b inset) Intraclique LLC/Shutterstock. (c inset) Radius Images/Alamy Images. (t inset) Peter Hansen/Shutterstock. **2-3** (bg) Fred Bavendam/Minden Pictures/National Geographic Image Collection. **3** (b inset) Jupiterimages/Getty Images. (t inset) Cyril Ruoso/Minden Pictures/National Geographic Image Collection. **4** (bg) Tom Schwabel/National Geographic Image Collection. (inset) Renee Fadiman.

Chapter 1
5 Specialist Stock/Corbis. **6-7** (ChapOp) Specialist Stock/Corbis. **8** (b) Bruno Morandi/age fotostock/Photolibrary. (c) Czintos Ödön/Shutterstock. (t) ISM/Phototake. **9** (b) Michael and Pataricia Fogden/Minden Pictures/National Geographic Image Collection. (c) Ron Steiner/Alamy Images. (t) Jonathan Blair/National Geographic Image Collection. **10-11** Michael and Patricia Fogden/Minden Pictures/National Geographic Image Collection. **11** (inset) ISM/Phototake. **12** (c inset) Dr. Dennis Kunkel Microscopy, Inc/Visuals Unlimited. (l inset) Dr. Terry Beveridge/Visuals Unlimited. (r inset) David M. Phillips/Photo Researchers, Inc. **13** (bg) Frank Krahmer/Corbis. (c inset) blickwinkel/Alamy Images. (l inset) Czintos Ödön/Shutterstock. (r inset) DigitalStock/Corbis. **14-15** Bruno Morandi/age fotostock/Photolibrary. **15** Row 1 L to R: 1 Vladimirs Koskins/Shutterstock. 2 PhotoDisc/Getty Images. 3 Bill Perry/Shutterstock. 4 James Pierce/Shutterstock. 5 PhotoDisc/Getty Images. 6 PhotoDisc/Getty Images. 7 G. K. & Vikki Hart/Photodisc/Getty Images. Row 2 L to R: 1 Vladimirs Koskins/Shutterstock. 2 PhotoDisc/Getty Images. 3 Bill Perry/Shutterstock. 4 James Pierce/Shutterstock. 5 PhotoDisc/Getty Images. 6 PhotoDisc/Getty Images. Row 3 L to R: 1 Vladimirs Koskins/Shutterstock. 2 PhotoDisc/Getty Images. 3 Bill Perry/Shutterstock. 4 James Pierce/Shutterstock. 5 PhotoDisc/Getty Images. Row 4 L to R: 1 Vladimirs Koskins/Shutterstock. 2 PhotoDisc/Getty Images. 3 Bill Perry/Shutterstock. 4 James Pierce/Shutterstock. Row 5 L to R: 1 Vladimirs Koskins/Shutterstock. 2 PhotoDisc/Getty Images. 3 Bill Perry/Shutterstock. Row 6 L to R: 1 Vladimirs Koskins/Shutterstock. 2 PhotoDisc/Getty Images. Row 7: 1 Vladimirs Koskins/Shutterstock. **16-17** Theo Allofs/Getty Images. **17** (inset) Stephen Dalton/Photo Researchers, Inc. **18-19** Jonathan Blair/National Geographic Image Collection. **19** (l inset) James Forte/National Geographic Image Collection. (r inset) David Hughes/iStockphoto. **20-21** DJ Mattaar/Shutterstock. **21** (l inset) Ron Steiner/Alamy Images. (r inset) Corel. **22** (b inset) Daryl H/Shutterstock. (t inset) Vinicius Ramalho Tupinamba/iStockphoto. **22-23** Fred Bavendam/Minden Pictures/National Geographic Image Collection. **23** (inset) Sergey Toronto/Shutterstock. **24-25** tbkmedia.de/Alamy Images. **25** (l inset) JH Pete Carmichael/Image Bank/Getty Images. (r inset) Konstantin Kalishko/Alamy Images. **26** (bg) John Anderson/Alamy Images. (inset) Will Rennick/iStockphoto. **27** (b) Joyce Gross Photography. (bc) NHPA/Photoshot. (c) Tom Tookey/Alamy Images. (t) klemens wolf/iStockphoto. (tc) Wildlife GmbH/Alamy Images. **28-29** WorldFoto/Alamy Images. **29** Patricia Danna/Animals Animals. **30** (inset) Chris Mattison; Frank Lane Picture Agency/Corbis. **30-31** Wild Wonders of Europe/Wothe/Nature Picture Library. **31** (l inset) Michael and Patricia Fogden/Minden Pictures/National Geographic Image Collection. (r inset) Joe Drivas/Iconica/Getty Images. **32-33** James Gerholdt/Photolibrary. **33** (inset) Peter Hansen/Shutterstock. **34-35** Michael and Pataricia Fogden/Minden Pictures/National Geographic Image Collection. **35** (inset) Daryl Balfour/Gallo Images/Alamy Images. **36-37** Juniors Bildarchiv/Alamy Images. **37** (inset) David Fleetham/Alamy Images. **38** Mark Thiessen/National GeographicSociety/National Geographic Image Collection. **39** (bg) Mark Thiessen/National Geographic Society/National Geographic Image Collection. (inset) Mark Thiessen/National Geographic Society/National Geographic Image Collection. **40-41** Sue Daly/ Nature Picture Library/Alamy Images. **41** (inset) Barry Mansell/Nature Picture Library. **42** Meg Lowman/Dr. Margaret D. Lowman. **43** (b) Mark Moffett/Minden Pictures/National Geographic Image Collection. (t inset) Chris Knight/Dr. Margaret D. Lowman. **44-45** Tim Flach/Stone/Getty Images. **45** (b inset) Oliver Lucanus/Minden Pictures. **46** (l inset) John Bell/Bruce Coleman/Photoshot. **46-47** tbkmedia.de/Alamy Images. **48** (b) Roy Toft/National Geographic Image Collection. **48-49** Watt Jim/Pacific Stock/Photolibrary. **50** (b) LYNN M. STONE/Nature Picture Library. **51** (b) cbimages/Alamy Images. **52** Watt Jim/Pacific Stock/Photolibrary.

Chapter 2
53 joSon/Getty Images. **54-55** joSon/Getty Images. **56** (b) DLILLC/Corbis Premium RF //Alamy Images. (t) Gallo Images/Corbis. **57** (b) Larry Lee Photography/Corbis. (c) Alex Wild/Visuals Unlimited. (t) Tom Walker/Visuals Unlimited. **58-59** Gallo Images/Corbis. **60** Chris Johns/National Geographic Image Collection. **62-63** (bg) Tom Walker/Visuals Unlimited. **63** (inset) Stephen Dalton/Minden Pictures. **64** (inset) Fritz Polking/Peter Arnold, Inc./Alamy Images. **64-65** (bg) Michael Obert/mauritius images GmbH/Alamy Images. **65** (inset) Juniors Bildarchiv/age fotostock. **66** Richard Du Toit/Minde Pictures/National Geographic Image Collection. **67** (bg) Bertram Murray Jr/Animals Animals. (inset) Alex Wild/Visuals Unlimited. **68** Masa Ushioda/Alamy Images. **69** George Grall/National Geographic Image Collection. **70-71** (bg) Larry Lee Photography/Corbis. **71** (l inset) Gerry Ellis/Minden Pictures/National Geographic Image Collection. (r inset) Kevin Ebi/Alamy Images. **72** (l inset) DLILLC/Corbis Premium RF/Alamy Images. (r inset) imagebroker/Alamy Images. **72-73** (bg) Michael Quinton./Minden Pictures/National

CREDITS

Geographic Image Collection. **73** (l inset) Stephen Saks/Photo Network/Alamy Images. (r inset) offiwent.com/Alamy Images. **74** (bg) Jason Edwards/National Geographic Stock/National Geographic Image Collection. (inset) john t. fowler/Alamy Images. **75** Peter Yates/Time Life Pictures/Getty Images. **76** (inset) Stone Nature Photography/Alamy Images. **76-77** (bg) Radius Images/Alamy Images. **77** (inset) SergioZ/Shutterstock. **78** geogphotos/Alamy Images. **79** (b inset) Alfred Eisenstaedt/Time & Life Pictures/Getty Images. (bg) Jeff Greenberg/PhotoEdit. (t inset) Alfred Eisenstaedt/Time & Life Pictures/Getty Images. **80** (inset) egd/Shutterstock. **80-81** (bg) Jorg Hackemann/Shutterstock. **81** (inset) Simon Fraser/Photo Researchers, Inc. **82** (inset) Aaron Haupt/Photo Researchers, Inc. **82-83** (bg) Mike Brinson/Getty Images. **83** (inset) Image Source/Getty Images. **84** (bg) Datacraft/Getty Images. (inset) Diane Cook and Len Jenshel/National Geographic Image Collection. **85** Diane Cook and Len Jenshel/National Geographic Image Collection. **86** John Shaw/Photoshot. **87** Alan and Sandy Carey/Getty Images. **88** (b) Luciana Brondizio/Milagros López Mendilaharsu. (t) Lisandro Almeida. **89** Projeto Tamar Image Data Bank. **90-91** Georgette Douwma/Getty Images. **91** (t inset) Jon Bertsch/Visuals Unlimited/Alamy Images (b inset) Brian J. Skerry/National Geographic Image Collection. **92** Oxford Scientific (OSF)/Photolibrary. **93** Rich Carey/Shutterstock. **94-95** (bg) Darlyne A. Murawski/National Geographic Image Collection. **96** (bg) Doug Perrine/SeaPics.com. (inset) Daniel Gotshall/Visuals Unlimited. **97** Debra James/Shutterstock. **98** Pixtal Images/Photolibrary. **99** (bg) Poelzer Wolfgang/Alamy Images. (inset) Poelzer Wolfgang/Alamy Images. **100** Georgette Douwma/Getty Images.

Chapter 3

101 Tim Laman/National Geographic Image Collection. **102-103** Tim Laman/National Geographic Image Collection. **104** ooyoo/iStockphoto. **105** (bg) BLOOMimage/Getty Images. (l inset) Brendan Bucy/Shutterstock. (r inset) Barry Mansell/Nature Picture Library. **106** (inset) Peter Baxter/Shutterstock. **106-107** (bg) Fred Leonero/Shutterstock. **107** (inset) Ben Acton/iStockphoto. **108** ooyoo/iStockphoto. **109** OSF/Milkins, C./Animals Animals. **110-111** Hans Christoph Kappel/Minden Pictures. **111** (inset) Michael Patrick O'Neill/Photo Researchers, Inc. **112-113** (bg) M. Watson/Ardea.com. **113** (inset) Breck P. Kent/Animals Animals. **114-115** Intraclique LLC/Shutterstock. **115** (inset) Eye of Science/Photo Researchers, Inc. **116** (c inset) Brendan Bucy/Shutterstock. (l inset) BLOOMimage/Getty Images. (r inset) Barry Mansell/Nature Picture Library. **116-117** (bg) Bob Gibbons/Alamy Images. **117** (l inset) Rusty Dodson/Shutterstock. (r inset) James McLaughlin/Alamy Images. **118-119** (bg) Jeff Foott/Getty Images. **120** (b) Emory Kristof/National Geographic Image Collection. (t) Emory Kristof/National Geographic Image Collection. **120-121** (bg) Emory Kristof/National Geographic Image Collection. **121** (inset) Emory Kristof/National Geographic Image Collection. **122-123** Garfield/Getty Images. **123** (c inset) Glen Allison/White/Photolibrary. (l inset) Annie Griffiths/National Geographic Image Collection. (r inset) Rod Planck/Photo Researchers, Inc. **124** Neo Martinez. **125** (bg) Rick Elkins/Getty Images. **125** (inset) Photography by Pamela Palma. **126-127** (bg) Ralph Lee Hopkins/National Geographic Image Collection. **127** (inset) Darlyne A. Murawski/National Geographic Image Collection. **128** (b) Ralph Lee Hopkins/National Geographic Image Collection. (cl inset) Flip Nicklin/Minden Pictures/National Geographic Image Collection. (cr inset) Wild Wonders of Europe/Lundgren/Nature Picture Library. (l inset) Manfred Kage/Peter Arnold, Inc. (r inset) Paul Nicklen/National Geographic Image Collection. **129** (bl) Bill Curtsinger/National Geographic Image Collection. (r) Jim Borrowman/All Canada Photos. (tl) Mark Peters/Alamy Images. **132** Darlyne A. Murawski/National Geographic Image Collection.

Chapter 4

133 Stuart Westmorland/Image Bank/Getty Images. **134-135** Stuart Westmorland/Image Bank/Getty Images. **136** (b) Michael Weber/imagebroker/Alamy Images. (c) ANT Photo Library/Photo Researchers, Inc. (t) Phil Degginger/Animals Animals. **137** (b) clickit/Shutterstock. (t) Mike McClure/Index Stock/age fotostock. **138-139** (bg) Jupiterimages/Comstock Images/Getty Images. **139** (b inset) Wildlife/Peter Arnold, Inc. (c inset) Volker Steger/Photo Researchers, Inc. (t inset) Bob Elsdale/Eureka/Photo Researchers, Inc. **140-141** (bg) Marcos Veiga/Alamy Images. **141** (bl) Paul Zahl/National Geographic Image Collection. (br) Christian Ziegler/National Geographic Image Collection. (tl) Darylne A. Murawski/National Geographic Image Collection. (tr) Michael and Patricia Fogden/Minden Pictures/National Geographic Image Collection. **142** (inset) Travelpix Ltd/Getty Images. **142-143** (bg) Justus de Cuveland/imagebroker/Alamy Images. **143** (inset) Wildlife/Peter Arnold, Inc. **144-145** Phil Degginger/Animals Animals. **145** Michael Weber/imagebroker/Alamy Images. **146** (bg) Cathy Keifer/Shutterstock. (inset) Creatas/Jupiterimages. **147** Mike McClure/Index Stock/age fotostock. **148-149** Cyril Ruoso/Minden Pictures/National Geographic Image Collection. **149** (b) ANT Photo Library/Photo Researchers, Inc. (c) Tom McHugh/Photo Researchers, Inc. (t) Elzbieta Sekowska/iStockphoto. **150** (bg) Dave Watts/Alamy Images. (inset) Michael Krabs/imagebroker/Alamy Images. **151** clickit/Shutterstock. **152** Brain J. Skerry/National Geographic Image Collection. **153** (b) Brian J. Skerry/National Geographic Image Collection. (t) Brian J. Skerry/National Geographic Image Collection. **154** ©George McCallum/SeaPics.com. **155** (b) Brian J. Skerry/National Geographic Image Collection. (t) Brian J. Skerry/National Geographic Image Collection. **156** (b) Purestock/Alamy Images. (inset) Skip Higgins of Raskal Photography/Alamy Images. **158** Ed Reschke/Peter Arnold, Inc./Alamy Images. **160-161** inga spence/Alamy Images. **161** (l inset) James Forte/National Geographic Image Collection. (r inset) J S Sira/Gap Photo/Visuals Unlimited. **162** (b inset) R B Forbes/ASM Mammal Image Library. (t inset) Rick & Nora Bowers/Alamy Images. **162-163** Phil Degginger/Alamy Images. **163** (l inset) FLPA/- David Hosking/age fotostock. (r inset) Mark Moffett/Minden Pictures/National Geographic Image Collection. **164** (bg) Lisa S. Engelbrecht / Danita Delimont.,/Alamy Images. (inset) Sykes, P.W./U.S. Fish & Wildlife Service. **165** David Fleetham/Visuals Unlimited. **166** (b) Bernardo Gonzalez Riga/epa/Corbis. (t) Kun Jiang/iStockphoto. **167** (bg) Volker Steger/Photo Researchers, Inc. (inset) Ken Lucas/Visuals Unlimited. **168** (l inset) Jupiterimages/Comstock Images/Getty Images. (r inset) Dave Watts/Alamy Images. **168-169** Digital Vision/Getty

EM17

CREDITS

Images. **169** (inset) Elzbieta Sekowska/iStockphoto. **170** Brant Allen. **171** (b) Sudeep Chandra/Zeb Hogan. (t) Courtesy of Zeb Hogan/Zeb Hogan. **172** (inset) Christian Ziegler/Minden Pictures/National Geographic Image Collection. (tr) Tomasz Zachariasz/iStockphoto. **172–173** Mark Moffett/ Minden Pictures/National Geographic Image Collection. **173** (tl) Tomasz Zachariasz/iStockphoto. **174** Mark Moffett/ Minden Pictures/National Geographic Image Collection. **175** (bg) Mitsuaki Iwago/Minden Pictures/National Geographic Image Collection. (inset) Martin Dohrn/Nature Picture Library. **177** (inset) Andrew Darrington/Alamy Images. **178** Mark Moffett/Minden Pictures/National Geographic Image Collection. **179** Dennis Kunkel Microscopy, Inc./Phototake/Alamy Images. **180** Mitsuaki Iwago/Minden Pictures/National Geographic Image Collection.

Chapter 5

181 ejwhite/Shutterstock. **182–183** ejwhite/Shutterstock. **184** (b) 81A Productions/Photolibrary. (c) Symphonie/Getty Images. (t) Laurence Monneret/Stone/Getty Images. **185** (b) Anders Rising/NordicPhotos/age fotostock. (c) Corbis. (t) Jupiterimages/Getty Images. **186–187** (bg) Laurence Monneret/Stone/Getty Images. **187** (inset) GJLP/Photo Researchers, Inc. **188–189** (bg) David Epperson/Photographer's Choice/Getty Images. **190–191** (bg) Cultura/Alamy Images. **192–193** (bg) Symphonie/Getty Images. **194–195** (bg) Martin Strmiska/Alamy Images. **196–197** (bg) 81A Productions/Photolibrary. **198–199** (bg) Jupiterimages/Getty Images. **200–201** (bg) Corbis. **202–203** (bg) Anders Rising/NordicPhotos/age fotostock. **204–205** (bg) Corbis. **205** (b) pjcross/Shutterstock. (bc) arenacreative/Shutterstock. (c) Stockbyte/Getty Images. (t) BananaStock/Jupiterimages. (tc) aleksandar zoric/Shutterstock. **206–207** (bg) H. Mark Weidman/Workbook Stock/Getty Images. **207** (b) Andreas Gradin/Shutterstock. (bc) Tim Laman/Nature Picture Library. (t) Stephen Frink Collection/Alamy Images. (tc) Bill Draker /Rolf Nussbaumer Photography/Alamy Images. **208** (b inset) Holly Kuchera/Shutterstock. (t inset) Anita Huszti/Shutterstock. **208–209** (bg) Creatas/Jupiterimages. **210–211** (bg) Tony Heald/Nature Picture Library. **211** (l inset) Michael Durham/Minden Pictures/National Geographic Image Collection. (r inset) GeoTravel/Alamy Images. **212** (inset) Vilmos Varga/Shutterstock. **212–213** (bg) Heidi and Hans-Jurgen Koch/Minden Pictures/National Geographic Image Collection. **213** (b) Jack Milchanowski/Visuals Unlimited. (t) Renaud Visage/Getty Images. **214–215** (bg) SMC Images/Getty Images. **215** (inset) Petronilo G. Dangoy Jr./Shutterstock. **216** Jon Gilbert Fox. **216–217** (t) Southern Stock/Brand X Pictures/Getty Images. **217** (bl) Dr. Lori Alvord. (br) Stock Connection Distribution/Alamy Images. **218** (inset) Reuters/Corbis. **218–219** (bg) Quinn Rooney/Getty Images. **220** Thomas Frey/imagebroker/Alamy Images. **221** Robert Oliver/epa/Corbis. **222** Mike Banks/Alamy Images. **224–225** Marius Becker/epa/Corbis. **226** (inset) Martinez de Cripan/epa/Corbis. **226–227** (bg) Hugh Gentry/Reuters/Landov. **227** (inset) Thomas Frey/imagebroker/Alamy Images. **228** Thomas Frey/imagebroker/Alamy Images.

End Matter

EM1 (bg) clickit/Shutterstock. (inset) ISM/Phototake. **EM2** (bg) John Anderson/Alamy Images. (inset) ANT Photo Library/Photo Researchers, Inc. **EM3** (l) Czintos Ödön/Shutterstock. **EM4** (bg) Bruno Morandi/age footstock/Photolibrary. (inset) Jonathan Blair/National Geographic Image Collection. **EM7** (t) Tomasz Zachariasz/iStockphoto. (b) Tomasz Zachariasz/iStockphoto. **EM9** Ralph Lee Hopkins/National Geographic Image Collection. **EM12** Ed Reschke/Peter Arnold, Inc./Alamy Images. **EM14–EM15** Juniors Bildarchiv/Alamy Images. **EM18** 81A Productions/Photolibrary. **Back Cover** (bg) Stephen Alvarez/National Geographic Image Collection. (bl) Georges Gobet/AFP/Getty Images. (br) Beverly Joubert/National Geographic Image Collection. (c) Alexander Joe/AFP/Getty Images. (tl) Maria Fadiman. (tr) Maria Fadiman.